Taking Back Our Lives in the Age of Corporate Dominance

Ellen Schwartz
Suzanne Stoddard

16
EasyRead Large

RHYW

Copyright Page from the Original Book

Taking Back Our Lives in the Age of Corporate Dominance

Berrett-Koehler Publishers, Inc.
235 Montgomery Street, Suite 650
San Francisco, California 94104-2916
Tel: (415) 288-0260, Fax: (415) 362-2512
www.bkconnection.com

Ordering information for print editions
Quantity sales. Special discounts are available on quantity purchases by corporations, associations, and others. For details, contact the "Special Sales Department" at the Berrett-Koehler address above.
Individual sales. Berrett-Koehler publications are available through most bookstores. They can also be ordered directly from Berrett-Koehler: Tel: (800) 929-2929; Fax: (802) 864-7626; www.bkconnection.com
Orders for college textbook/course adoption use. Please contact Berrett-Koehler: Tel: (800) 929-2929; Fax: (802) 864-7626.
Orders by U.S. trade bookstores and wholesalers. Please contact Ingram Publisher Services, Tel: (800) 509-4887; Fax: (800) 838-1149; E-mail: customer.service@ingrampublisherservices.com; or visit www.ingrampublisherservices.com/Ordering for details about electronic ordering.

First Edition
Paperback print edition ISBN 978-1-57675-078-0
PDF e-book ISBN 978-1-60994-161-1
IDPF e-book ISBN 978-1-60994-604-3

2011-1

Designed by Detta Penna

This optimized ReadHowYouWant edition contains the complete, unabridged text of the original publisher's edition. Other aspects of the book may vary from the original edition.

Set in 16 pt. Verdana

TABLE OF CONTENTS

"Taking Back Our Lives is a powerful examination of the forces we all face in a society pulling us away from our authentic, deeper selves. It offers refreshing solutions that speak to the heart and uplift our spirits."

—Scott and Shannon Peck,
coauthors of *Liberating Your Magnificence:*
25 Keys to Loving and Healing Yourself

"Taking Back Our Lives in the Age of Corporate Dominance embodies the message of conscious evolution by giving us guidance on how to transform the corporate structure."

—Barbara Marx Hubbard,
author of *The Revelation:*
A Message of Hope for the New Millenium

"Taking Back Our Lives is a very courageous book by two highly qualified women who dare to speak their truth about a reality the media avoids—the corporate consumer culture eroding our democracy—and then show us practical alternatives that start right at home. A timely imperative for survival. Read it and act now!"

—Elisabet Sahtouris,
author of *A Walk Through Time*
and *Biology Revisioned*

"***Taking Back Our Lives*** is not just a book. It is a bold and courageous exposé of what is happening to our society without our conscious knowledge. It is well-documented and thought-provoking, giving substance to what most of us only talk about among ourselves. It challenges all Americans to wake up, offering positive action steps that each of us can take to empower ourselves individually and collectively."

—Rama Vernon,
Founder, Women of Vision and Action,
and President, The Center for International Dialogue

"***Taking Back Our Lives in the Age of Corporate Dominance*** is a welcome blueprint we should not only read, but also live by. My life changed by reading and living the message."

—Howard F. Lyman, author of *Mad Cowboy*

"This book is more than timely. It is a vital guide to the self-improvement we need to move off of the self-destructive path that threatens us and our future."

—Margaret Wheatley,
author of *Leadership and the New Science*
and coauthor of *A Simpler Way*

"Schwartz and Stoddard have the best plan yet to deal with the world increasingly run by corporations and money: personal disengagement. They tell us how to find our own 'moral alternative' to economic globalization and live it. A great book."

—Maude Barlow,
Volunteer National Chairperson of the Council of Canadians

"***Taking Back Our Lives in the Age of Corporate Dominance*** interlaces the personal and the political. An original combination of critical politics and New Age optimism."

—Michael Parenti,
author of *History as Mystery* and *America Besieged*

"***Taking Back Our Lives*** is a powerful and sharp-cutting truthtelling, a necessary awakening, and an extraordinarily practical collection of solutions for *literally* the survival of all life on earth."

—Thom Hartmann,
author of *The Last Hours of Ancient Sunlight* and *The Prophet's Way*

I dedicate this book to my daughter, Andrea, who gave up countless hours with me so that I might speak my truth, and to my father John, who is on the other side. He believed in me and was proud of me before I had any inkling who I was.
—Ellen

I dedicate this book to my brother Rupert, who died well before I ever thought of writing a book, and who remains my greatest inspiration in giving and forgiving.
—Suzanne

We dedicate this book to all the children of the world who are counting on each of us to make the choices that will create the world anew.

Foreword

Socrates reminded us almost 2400 years ago that "the unexamined life is not worth living." Even though the examination is essential, it can be uncomfortable. Socrates himself found out how true this could be. When it seems that things are "on a roll," we don't always want to ask questions that might expose any undesirable consequences.

There is no denying that the technological advances made by large corporations in the twentieth century have brought many benefits to all of us. Diseases that used to kill millions of people have been almost eradicated. Modern transportation not only makes travel easy and comfortable, but it can also bring help within hours when disaster strikes anywhere in the world. Media developments like the Internet, digital TV, and electronic games provide entertainment and information while linking billions of people globally. Many of us who enjoy these developments do not want to confront the troubling questions about whether this progress comes with a price.

To ask the difficult questions does not negate the positives. In fact, the *examined* life can turn out to be much more worth living. It is helpful, of course, if the problems that are uncovered are accompanied by solutions. The book you are about to read accomplishes this very difficult task.

Taking Back Our Lives challenges and inspires. The authors present information about the modern corporate world that is deeply troubling. But they don't stop there. These two women combine the hard-hitting facts with readily available tools for reclaiming our democracy, our work lives, our children, and our time. *Taking Back Our Lives* is a guidebook for finding and creating meaning at home and at work. Ellen Schwartz and Suzanne Stoddard remind us there is no one out there who will do it for us. We are the ones we've been waiting for. The 75 action items for inner exploration and effective actions in the outer world light the way.

The two women who wrote this book have amassed solid information from their years of research. But they also learned many of these lessons firsthand. Fortunately for us, they are not reluctant to share their journeys—the mistakes along with the discoveries. The result is that we get a sense that we are not alone and that there is a way to maximize the benefits of modern society while minimizing the harm.

We sometimes forget that the real meaning of responsibility is the "ability to respond." The only way for society to improve is for individual members to stand up, speak out, and offer constructive suggestions. I, for one, am grateful that Ellen Schwartz and Suzanne Stoddard have done that. After reading *Taking Back Our Lives,* I am confident that you will be too.

David Walsh, Ph.D.
President, National Institute on Media and the Family

and author, *Selling Out America's Children*

Acknowledgments

From Ellen

Appreciation goes to Michael Schwartz, who encouraged me to sit down and start writing when this book was but a distant dream. Suzanne joins me in thanking Michael for the (generally!) good humor he showed, even as our writing project took over the living room and kitchen for many months. Heartfelt thanks also go to my kindred spirit Gayle Curtis, who understands beyond words and replenishes my spirit across mountains and deserts.

From Suzanne

My husband Ruel Robbins deserves thanks for putting up with all my absences from home while I stayed at the Schwartz's to write. His consistent support and love over the 27 years of our relationship have given me roots and wings. I am most grateful to my mother Sally Stoddard for her thoughtful feedback, fresh viewpoints, title suggestions, and open invitation to take nurturing breaks in Palo Alto. She also has my thanks for helping with the Bibliography and Resource lists. I am very blessed that my father Richard Stoddard instilled in me a love of language and words, and that my brothers Rick and Kirk have never failed to offer loving encouragement. Thank you also to my dearest friend, Sara Tarr, for continually cheering me on.

From both of us

We wish to acknowledge Mary and Phil Strauss for their kindness in letting us stay, at the project's inception, in Mary's childhood home in Carmel, one of the loveliest places on earth. Thanks also to Sergio and Gaye Lub for graciously allowing us to use their guest home in Napa so that we might have the freedom to write day and night as the first draft drew to completion. Sergio's abiding interest in our project manifested itself in many helpful ways, including his careful reading of our manuscript and the networking he did on our behalf.

And, finally, a big thank you to Berrett-Koehler's Steve Piersanti and Valerie Barth for their belief in us, for their insightful critiques, and for the freedom to participate in every aspect of the publishing process. Their contract must be the most enlightened and author-friendly of any publisher's. Everyone we've interacted with on their staff has been accommodating, highly professional, and a joy to work with.

Part I

How Corporate Structures, Products, and Processes Impact Our Work and Personal Lives

Introduction

Suzanne Stoddard

a snapshot of the pressures and potentialities of our consuming corporate times

The lust for comfort murders the passion of the soul,
and then walks grinning in the funeral.
—Kahlil Gibran

This book is for those who wish to find out who they are and what they are going to do about it while they still have time.

If we lived in a healthier culture, a book like this wouldn't be necessary. In a saner society, all of us would have an opportunity to contribute in ways that reflected our unique talents, and we would take much better care of each other. We would allow plenty of time to dance, to make music, to create, to play, and to revel in being human. We would spend most of our time doing what makes us feel alive and laugh as often as happy children.

Instead, we feel stressed by the relentless demands of seeking or holding a job and from the inability to find enough time for ourselves, our families, and our communities. In our heart of hearts, we know there is a more enticing life out there, but we don't know how to find it, and we don't even have enough time to look!

We have made *comfort* a premier value and created enough distractions to suppress *consciousness* forever. Currently, most adults spend fifteen to twenty hours a week watching other people have experiences—in the television, video, and virtual media-rather than having their own. Most children spend twice that amount of time in the spectator realm. Because working and commuting take so much of our time and life energy, we have lost the will to spend our free hours in truly experiential activities, and to share them with the young people in our lives. But the glut of entertainment bores and numbs us. Like all addictions, it takes from us far more than it gives. And the craving to feel good all the time keeps us from feeling alive.

We buy homes that require two wage-earners to cover the mortgage. Thanks to the slick advertising that surrounds us on television, radio, billboards, newspapers, magazines, and the Internet, our desires are teased far beyond the time or money available to satisfy them. The continual distractions of our highly competitive culture estrange us from our actual needs and natures. We've forgotten that joy comes from meaningful work, loving relationships, creative pursuits, personal growth, service to others, mental and physical exercise, and recreation that actually re-creates us.

Because the major media present information in a segmented and superficial way, we often feel powerless to make a difference in our own communities. It is rare for newscasters to note the cause-and-effect links between the bottom-line-driven practices of transnational companies and the loss of living-wage jobs, the capacity to protect our environment, and growing substance abuse across all age groups. Indeed, corporations have unprecedented power today, impacting far more than our paychecks and the products we buy with them.

Certainly, corporations have brought great benefits to millions of people, raising the standard of living in America to the highest in the world. Yet there have been profound unforeseen consequences that are now degrading the quality of our lives. The middle class is joining the poor as the "anxious class." Even those who are comfortable today fear they could lose it all tomorrow. Insecurity breeds stress and puts us all on a short fuse. And the mountains of soul-killing paperwork that must be dealt with daily in order to avoid chaos are a major challenge for even the well educated and well organized.

In Part I, we present the darker side of transnational corporate growth that is rarely depicted in the mainstream media. We begin by examining the corporate infrastructure, processes, and imperatives, which are the underpinnings of our technological culture. We penetrate the many myths surrounding competition, probing its deeper impact on our work and personal lives. The need for relief from stress draws us to the easy escape of television. Yet the entertainment industry is both a conveyor and a major instigator of increasing disrespect and alienation, as well as diminished creativity and problem-solving ability among our youth. We detail the connections between the epidemic of eating disorders among our girls and the impossible standard of thinness marketed in virtually all media.

We take you directly into a Congressional campaign where carefully crafted imagery and strategically placed corporate money drive the results. We do a reality check on globalization, showing how trade treaties underwritten by transnationals promote their own highest profits at the expense of the Earth herself and of all living things. As corporations have grown from national businesses to international marketers to transnational operations—above the reach of the law of the countries in which they operate—the pace of our lives has accelerated and our very biological rhythms have been disturbed. Time is a nonrenewable resource, and taking it back has become a necessity.

Part II offers simple tools, immediate actions, and larger strategies for reclaiming our lives in a world dangerously out of balance. Into every chapter we weave personal stories about confronting change, fear, risk, anger, inertia, and family conflicts. The inner clarity that comes from slowing down, reflection, and turning away from the numerous distractions of our culture is the springboard for meaningful action in our personal and economic lives.

Change is difficult. Though we may be in pain, we rarely welcome change. Ellen's story of being left in a Tucson boarding school because of her life-threatening asthma shows that even traumatic change can be transformative. Change can also be initiated from tapping into ancient wisdom traditions that move us away from our typically analytical approaches into nonlinear practices. Meditation, chanting, dance, and journaling soothe our spirits and plumb our depths.

Opportunities for personal evolution are generally close at hand, right in the family crucible. Those nearest us attune to our intentions as well as our actions. The situations that ignite our anger also hold the power to illuminate parts of ourselves that are aching for actualization. Because the hectic nature of modern life makes it difficult even to schedule meals together, we suggest many ways to enrich and soften family time.

But no matter how hard we try to make our family time nurturing, if we are not satisfied at work the residue taints our hours at home. Finding work that feeds our soul as well as body is no small task, yet the possibilities have never been greater. We are privileged to live in a time of paradigm shift, when the potential exists for transitioning from "jobs" to "true work" capable of healing the planet and bringing forth our deepest creative powers. The call to help humanity and the Earth can also lead us into service that offers great emotional richness.

As we forge new paths through our stressful, hi-tech world, the need for supportive associations is more important than ever. People are finding ways to relieve isolation and create community through potluck suppers, discussion circles, e-mail exchanges, and cohousing. The more real our relationships with those close to us, the fewer gaps we need to fill with expensive purchases that spend down our resources. If two heads are better than one, even more powerful are two hearts. There is synergy in sharing and collaboration. This book is a perfect example of that. Although you will see only one of our names at the top of most chapters, the book is ours together. In chapters with personal stories, we have indicated primary authorship so you will know who is speaking, but our words and ideas thoroughly entwined through the months of writing.

In offering this book, we *do* have a bias. We believe that the most enthusiastic and fulfilled people are those for whom life is a verb. These individuals continually find ways to learn and take risks, help and support, teach and guide, create and inspire, engage and enjoy. Because they know how to simplify and focus their own lives, they are in perpetual self-development and have a kind of fearlessness about testing boundaries. We believe that everyone is capable of living this way if they choose and that it's never too late to change your life course. Further, we believe that the survival of our planet depends on each one of us developing that part of our nature that seeks to love and contribute to the well-being of others. Finally, we believe the planet's survival depends on our paring down our material needs and simplifying our lives.

This book will stir up deep feelings and strong reactions. Some of the things we say will seem like an attack on your own particular sacred cows. (We know how you feel. We don't like being told that coffee—our drug of choice—isn't good for us.) But please don't put the book down when you come to one of these. Bear with us. Look at what we've written as a provocation that can help you see things in a different light, question old assumptions, and get your juices flowing. If we only seem to present one side, it is because we feel the other side gets plenty of coverage in the major media. Let our comments fuel your interactions with competition, the entertainment industry, Barbie, and the corporate world.

This book is about both outer action and inner growth. It is subversive. It is about *taking time to live* in a society increasingly dominated by mega-corporations that prefer we *not* follow our inner promptings. For, the more satisfied we are with our lives, the less we need what they have to sell. The more we esteem ourselves, the less we need expensive cars, designer clothing, and credit card getaways. The more time we take to go deep and discover our own beliefs, the less susceptible we are to the addictions and compulsions seen as social norms today. The power to choose is an incredible gift we have always had. In fact, *choice is our most important tool.* We can begin right now to use choice more consciously for our own and others' benefit.

Because awareness without action leads to frustration, we offer a wide array of possible steps—small and large, immediate and long-term—to choose from. We include *Action Items* in boxes within each chapter as well as *Questions for Reflection,* and *Resources* at the back of the book by chapter, so you can explore the themes and issues that engage, outrage, or inspire you. Resources and initiatives toward new ways of thinking, doing, working, and being are sprouting up everywhere. It is impossible to create an exhaustive, up-to-date list of everything available, but once you make a connection with any of the resources, you will be led to others or think of more yourself as you create your own new directions.

These are extraordinary times in which to be seeking our authentic selves.

Chapter 1

A Time Of Turbulence

Ellen Schwartz

when too much information keeps us from knowing the truth

If families just let the culture happen to them, they end up fat, addicted, broke, with a house full of junk and no time.
—Mary Pipher

You can never get enough of what you don't really want.
—Eric Hoffer

We must do the things we think we cannot do.
—Eleanor Roosevelt

We are overstimulated and disconnected from ourselves and nature. Image, not substance, is the stuff of our lives: how we look, what we wear, the car we drive. Even our democracy is driven by appearance. Candidates are packaged and rehearsed to speak in ninesecond sound bites.

We live in the Information Age. Translation: *too much* information, *too little* meaning, and *too little* wisdom. When Thich Nhat Hahn, the Vietnamese monk nominated for the Nobel Peace Prize, came to America, he said "Here I am in the land of the Information Superhighway. But do you know what one of your greatest problems is? Communication between each other!"

Not only is communication a problem, but rudeness and disrespect are rampant. Political exchanges are shouted. Talk-show hosts like Jerry Springer and Howard Stern bring new meaning to the word *offensive.* Verbal abuse abounds in movies and television shows that pass themselves off as entertainment. Comedy shows regularly descend to put-downs underscored by a laugh track. Teachers are noticing that many children don't know how to develop friendships; insults don't foster trust.

We are so cut off—heart from mind, feelings from action—that we do not respond with appropriate outrage at the "entertainment" that inundates our children. The interactive video games are the worst, because the children themselves take part in the murdering and maiming.

"Family values" is a political buzz word, but in reality, our political and economic systems do little to support families. Despite an increasing number of jobs, we have an actual loss of jobs that pay a living wage. While the defense budget catapults above and beyond what even Pentagon brass request, there is less money available to renew our schools for the twenty-first century. As the public airwaves are jammed with messages to buy, enjoy, and indulge, we are more often called consumers than citizens. Our familial and societal fabric is so frayed that more than 38,000 people die from guns each year, over half from arguments, accidents, and suicides. To try to ease our pain, to find a place of refuge, we are offered not only the malling of America but the walling of America.

What is it we're missing? Perhaps listening. Stories. Actions beyond our own lives.

Listening. Listening deeply. This is not something we are accustomed to in the era of the John McLaughlin model of communication. Too often we find ourselves jumping down the other person's throat before they even have a chance to finish their thought. This does not make them feel heard, appreciated, or understood.

Try listening to someone without interrupting or preparing a rebuttal in your mind while they're talking.

Stories. We can barely avoid hearing the stories of the rich and infamous. We know the polished details, spun to make their sex lives, love, rage, or pain much bigger than ours. We know those stories better than those of our own families or neighbors. Yet, it is the stories of the real people connected to us that ultimately hold meaning for us. What a loss it is, when families gather for the holidays, to have half the family bellowing for strangers running after a ball, while the heartbreaks and hopes, tough times and triumphs of those closest to them go unexpressed.

Invite a family member to sit down for a cup of tea. Tell them you'd like to get to know them on different levels. Ask some questions:

- *If you could have an hour with someone you admire, living or historical, who would that be?*
- *What was something you initially didn't want to face that you came through and felt good about later?*
- *If you died tomorrow, what would you want to be remembered for?*

Actions beyond our own lives. Though this is the land of rugged individualism, and independence is what we are taught to exalt, it is really interdependence which allows us to survive and flourish. Ralph Nader said "If we don't spend enough time in our public citizen lives, we will never have truly happy private citizen lives." We must work on several levels at once, moving fluidly in and out of the personal, familial, societal, political, and corporate arenas.

For too long we have fallen prey to the expert syndrome. We are supposed to be experts before venturing an opinion in a particular field. Most people don't paint or draw after the primary grades, because they're not "artists." Most don't enter into spirited discussions on the economy, because they're not economists. And we're not supposed to object to all the money still going into nuclear weapons, because we're not defense analysts. Yet, all it takes is common sense to see what's happening to our lives and to the planet.

Stress and hurry are the hallmarks of our time. Yet, the core of life is simple. We are called to take care of ourselves, keep our bodies healthy, and develop our own unique gifts with zest—even passion. We are called to spend time with those we love, being kindly and helpful, taking care not to discourage or humiliate. And we are called to do something in the broader community to make the world a better place than we found it.

This is neither easy nor hard. It merely involves taking a small step each day. Nothing grandiose. Just consistent actions.

The Native American four-fold path, as taught by Angeles Arrien, has a lot to offer us in these times. The first step is to *show up:* be where you need to be when you need to be there. The second step is to *be aware:* identify as much as possible what you are bringing to the situation—thoughts, emotions, assumptions, judgments, expectations. Tune in to the environment, too, and to what your and other peoples' physical condition and state of mind seem to be. Third, *speak your truth without blame or judgment.* Your truth is simply that: the consummation of *your own* thoughts and experiences. It is not eternal truth, or the "right" or "only" truth. Offer it as a perspective that may help others decide what is most appropriate for them at that moment. Fourth, *let go of outcome.* Perhaps this is the most difficult. We live under the illusion not only that we can control others but also that it is good to manipulate a situation to our own ends. Yet, we cannot really control anyone but ourselves (and often we can't even do that!).

For too long we've emphasized freedom in this country and forgotten responsibility. We've organized our time and our lives around the pleasure principle, instead of around true happiness and joyful service.

Experiment with the Native American fourfold path in three situations this week.

In these times, we have the privilege of awakening. While we have been anesthetized watching "the good life" on TV, fear, anger, and alienation have grown too strong in the land. Yet there is room for optimism. Getting in touch with the deep silent spaces within ourselves, coming together with hearts and minds, we can change the face of our communities, our nation, and the world.

Questions For Reflection

- What part of modern life is most raw to you?
- Do you feel deeply listened to? By whom?
- What stories do you hunger to know? To share?
- What part does selfless service play in your life?

Chapter 2

The Gift That Keeps On Taking

Ellen Schwartz

how the bottom-line mentality is bottoming out our lives and the planet

Economic globalization is shifting power away from governments responsible for the public good and toward a handful of corporations and financial institutions driven by a single imperative—the quest for short-term financial gain. This has concentrated massive economic and political power in the hands of an elite few ... Faced with pressures to produce greater short-term returns, the world's largest corporations are downsizing to shed people and functions.... It is becoming increasingly difficult for corporate managers to manage in the public interest, no matter how strong their moral values and commitment.

—David Korten, *When Corporations Rule the World*

Growth for the sake of growth is the ideology of the cancer cell.

—Edward Abbey

There is no polite way to say that business is destroying the world.
—Paul Hawken, *The Ecology of Commerce*

There is enough for everybody's need, but not for anybody's greed.
—Mahatma Gandhi

A house un-american activities committee today might well pose the question: "Are you now, or have you ever been, a disparager of corporations and consumerism?"

Our American Way of Life inseparably fuses with corporations and consumerism, and neither the media nor our political "leaders" are willing to take on the dark side of these powerful, omnipresent forces. We see occasional articles that hit the mark but little aggressive or sustained criticism. In fact, the only place we find "family values" in corporate America is on television commercials for laundry detergents and long distance phone service.

How has this come to be, and what does it all mean? What is the corporate framework of our technological society doing to us?

How does it impact our bodies and spirits to live in a country where every town is starting to look the same, with its Office Max and Citibank, McDonald's and Exxon? To live in houses in suburbs where we have to get in the car and burn gas to buy a loaf of bread or a book? To know that our natural environment is being chopped down and paved over as fast as developers can get to it? To be working longer hours and still not be able to keep up with the bills? To be afraid of the night, the dark and the sounds, because of the random and brutal crime images that pervade evening newscasts?

Can Corporations Be Compassionate?

Corporate products, processes, and structures impinge upon almost every aspect of our existence. Many of us think that if only more compassionate and broad-minded people could be in positions of top management, corporations would be more responsive and responsible. However, Jerry Mander points out in his leading edge book, *In the Absence of the Sacred,* that management must follow corporate law:

> U.S. corporate law holds that management of publicly held companies must act primarily in the economic interests of shareholders. If not, management can be sued by shareholders and firings would surely occur. So managers are legally obliged to ignore community welfare (e.g., worker health and satisfaction, environmental concerns) if those needs interfere with profitability. And corporate managers must also deny that corporate acts have a negative impact of any kind, if that impact might translate into costly damage suits that hinder profits.[1]

Does this seem too extreme to be true? Do you remember Bhopal? Mander recaps the story:

> In 1986, Union Carbide Corporation's chemical plant in Bhopal, India accidentally released methyl isocynate into the air, injuring some 200,000 people and killing more than 2,000.

Soon after the accident the chairman of the board of Union Carbide, Warren M. Anderson, was so upset at what happened that he informed the media that he would spend the rest of his life attempting to correct the problems his company had caused and to make amends. Only one year later, however, Mr. Anderson was quoted in *Business Week* as saying that he had "overreacted," and was now prepared to lead the company in the legal fight *against* paying damages and reparations. What happened? Very simply, Mr. Anderson at first reacted as a human being. Later, he realized (and perhaps was pressed to realize) that this reaction was inappropriate for a chairman of the board of a company whose primary obligations are not to the poor victims of Bhopal, but to shareholders; that is, to its profit picture. If Mr. Anderson had persisted in expressing his personal feelings or acknowledging the company's culpability, he certainly would have been fired.[2]

Join others in efforts to hold managers and stockholders accountable for harms, injuries, and debt. Call the International Forum on Globalization in San Francisco at 415-771-3394.

CEO Greed and the Declining Standard of Living

Mander continues, in *The Case Against the Global Economy,* "Corporate ideology, corporate priorities, corporate styles of behavior, corporate value systems, and corporate modes of organization have become synonymous with our way of life."[3]

What does this mean? What are the connections and consequences that are never broached by the talking heads on TV? The growth of poverty in our midst and the increasingly difficult time the middle class is having making ends meet are a direct effect of the astronomical increases in CEO compensation and shareholder gains, as well as politically influenced changes in the tax code.

In 1978, CEOs of large corporations made approximately 60 times more pay than their average worker. By 1997, that ratio had risen to 189:1. The 1998-99 version of *The State of Working America* by Lawrence Mishel, Jared Bernstein, and John Schmitt, confirms the lowering of living standards for the majority of Americans. For most people, compensation (wages plus benefits) fell between 1989 and 1997, with men experiencing a 7.8% drop. Looking at wages alone, between 1989 and 1997, the average hourly wage for men with a high school diploma and one to five years of work experience fell 7.4%; for comparable women, real wages fell 6.1%.[4]

Intel Corporation's chairman of the board and former CEO, Andrew Grove, made $97 million in 1996—3,528 times what their average factory worker makes! Excessive salaries are deductible as legitimate business expenses and vastly reduce the amount of taxes a corporation pays! Rep. Martin Sabo is a leader in introducing legislation that would amend the Internal Revenue Code to deny companies a deduction for payments of excessive compensation. Call his office at 202-225-4755 to find out the status of the Income Equity Act and encourage your own representative to co-sponsor it. You can also find this information from United For a Fair Economy *at* www.stw.org*or 617-423-2148.*

Much of the conventional economic analysis argues that a college degree is a prerequisite to participating in the new technology-driven economy. Yet the hourly wages of entry-level college graduates fell about 7% for both males and females between 1989 and 1997. Even newly hired engineers and scientists earned, respectively, 11% and 8% less than their counterparts in 1989 (adjusted for inflation). Mathematics and computer science graduates made modest gains, earning 5% more than the 1989 wage scale.

Looking at actual wages paid across the board, 28.6% of American workers earn less than $7.79 per hour—the amount it takes to lift a family of four above the poverty line with full-time, full-year employment. An additional 14.4% earn between $7.80 and $9.99 per hour. Only 57% of the American workforce earns $10 or more per hour.[5]

Working people got a slight reprieve in 1998 when, at last, real wages grew 2.6% for the typical (50th percentile) worker. This gain was not simply the result of increased wages; the typical family worked 129 more hours than in 1989—the equivalent of 3 weeks of full-time work to achieve this boost. This while their labors were generating a 9% increase in productivity for their companies.[6]

Did the productivity gain bring an increase in benefits? Not at all. Health insurance for those working in the private sector *decreased* from 70% in 1979 to 64.2% in 1997, while pension coverage went down from 51% to 47%.[7]

> *In 1955, union membership was 33%. In 1998, it was 14%. Unions were on the forefront in bringing about labor rights such as the 40-hour work week, vacation pay, healthcare benefits, pension packages, and safe working conditions. With increases in union-busting strategies following President Reagan's breaking of the air traffic controllers' strike in 1981, taken-forgranted benefits are declining. Write a Letter to the Editor in support of a local strike.*

Of course, the headlines in the business pages of the newspapers during the 1980s and 1990s were loudly proclaiming the good news of the stock market boom. The market's wildly rising prices, however, had little impact on working families, for the simple reason that the broad middle class does not own much stock. Almost 90% of all stock is owned by the wealthiest 10% of households; 60% of Americans own no stock at all. Eighty-six percent of the benefit of the 1989-1997 stock market increase went to the richest 10% of American households.[8]

Additionally, changes in the tax structure have shifted more burden to the middle class. In the 1950s, corporations paid 39% of the tax base; in the 1990s they contribute 10%. The theory behind giving corporations tax breaks is that they will create more jobs. Yet Fortune 500 companies eliminated 4.4 million jobs between 1980 and 1993, with the ten largest corporations alone shedding almost 500,000 jobs since 1990.[9] Corporate downsizing hit an all-time high in 1998, cutting 678,000 jobs in one year alone. The record $1.2 trillion in mergers fueled the layoff total, accounting for more than 10% of lost jobs, and merger mania continues.[10]

As newly laid-off workers wonder how they're going to pay next month's rent or mortgage, CEOs benefit financially from the job cuts. In a report by the Institute for Policy Studies, "Executive Excess: CEOs Gain from Massive Downsizing," Anderson,

Bayard, Collins, and Cavanaugh found that in the sixteen largest downsizers of 1998, all but three of the CEOs were rewarded with increases in salary, bonuses, and stock options. "On average, they received increases of 20% to $2.5 million—even higher than the average of $2.2 million for top executives surveyed by *Business Week.*" In the light of their findings, the authors propose a freeze of CEO pay during periods of downsizing.

Familiarize yourself with the latest CEO windfalls. Subscribe to the newsletter, Too Much: A Quarterly Commentary on Capping Excessive Income and Wealth. Call 1-800-316-2739.

The future looks even bleaker for job loss. Automation is the main reason. Jeremy Rifkin points out, in *The End of Work:*

> Current surveys show that fewer than 5% of companies around the world have even begun the transition to the new machine culture. This means that massive unemployment of a kind never before experienced is all but inevitable in the coming decades.... In the auto industry, each robot replaces four workers, and if in use twenty-four hours a day, will pay for itself in a little over one year.... Nippon Steel's new plant near Gary, Indiana has reduced the production cycle of some items from twelve days to one hour.... General

> Electric, a world leader in electronic manufacturing, has reduced employment from 400,000 in 1981 to less than 230,000 in 1993, while tripling its sales.[11]

It is estimated that 90% of customers will be using automated teller machines by the year 2000. The next time you use your ATM card, think about the teller jobs that are evaporating.

Who Benefits?

It is a common misperception that the transnational corporations are the major job providers. Although the Global Top 200 control more than a quarter of the world's economic activity, they employ less than 18.8 million people, which is less than a third of 1/100th of 1% of the world's workers.[12]

Wal-Mart, the twelfth largest corporation globally, is the typical new employer. A study by Kai Mander and Alex Boston reveals that the world's largest retailer, which opens a new store every three days, had profits of $2.6 billion in 1993 alone.

> Sam Walton, the founder of Wal-Mart, was unfailingly portrayed by the media as a folksy, kind-hearted embodiment of the American Dream.... In reality, Mr. Sam was in the business of driving small shops out of business and paying most of his workers minimum wage, while he amassed a personal fortune of $23.5 billion.... Most employees are given only part-time work

> so that the company can avoid paying the benefits full-time workers must receive.[13]

Even the low prices are a myth:

> Wal-Mart puts its fifteen hundred or so lowest-priced items at the front of the store. These are everyday products such as toothpaste and toilet paper of which people tend to know the price. Deeper into the store are the approximately eighty thousand high-profit items. As the local competition decreases, all of Wal-Mart's prices begin to climb.[14]

Relentless Corporate Brainwashing

While corporations would like us to believe that their efforts are ultimately for our benefit, the reality is much different. To maintain their image and promote their products and services, corporations spend $150 billion per year on advertising. The 22,000 advertisements we see each year are crafted by people whose job it is to make us unhappy with what we have and who we are. Their techniques work well—by and large we do believe that more is better, and new is always improved. Yet when we get the products, very often the satisfaction is not long-lasting. Consider, too, the time-saving devices. Do any of us have more leisure since the dawn of the "timesaving" era? We're working longer hours to buy, repair, and insure our possessions. Then there's the need for a house big enough to store it all.

Perhaps we might be happier if we pondered Mother Teresa's question to Americans: "What is enough?" Do we really need so many clothes, especially with the growing knowledge that most are made in sweatshop conditions? Must we have all the cheap plastic toys from China to fill party favor bags that hardly make it home without breaking? We in America are a large part of the 20% of the world's people consuming 80% of the earth's resources, mostly for making non-essential products.

The GDP: An Inadequate Measure of Economic Health

On paper it appears that the more products made, the better. The current economic indicator, the Gross Domestic Product (GDP), measures every monetary transaction. Economists Ted Halstead and Clifford Cobb point out that negative events such as depletion of natural resources, construction of more prisons, manufacture of bombs, and divorce that creates two households are all measures of "health" by this economic indicator.

> By itself, the GDP tells very little. Simply a measure of total output (the dollar value of finished goods and services), it assumes that everything produced is by definition "good." It does not distinguish between costs and benefits, between productive and destructive activities, or between sustainable and unsustainable ones.... By the cu-

> rious standard of the GDP, the nation's economic hero is a terminal cancer patient who is going through a costly divorce. The happiest event is an earthquake or a hurricane. The most desirable habitat is a multi-billion dollar Superfund site. All these add to the GDP, because they cause money to change hands. It is as if a business kept a balance sheet by merely adding up all "transactions," without distinguishing between income and expenses or between assets and liabilities."[15]

Write an article for the newsletter of an organization you belong to about the dangerously imbalanced picture the GDP presents by ignoring ecological losses in the equation.

The GDP was never meant to be an overall indicator. It was created during the Depression and came to its present form in World War II when production was all that mattered.

A Better Economic Indicator

Halstead and Cobb have come up with an ingenious new measurement: the Genuine Product Indicator (GPI). The GPI has both credit and debit columns. It subtracts social ills and ecological losses, and acknowledges unpaid household work, childcare, community service, and home growing of food.

Another crucial difference between the two indicators is that the GDP takes no account of income distribution. Over the past 20 years, the GDP rose 55%, yet real wages dropped by 14%. The top 5% of households enjoyed an income boost of 20%. One percent of Americans (those worth $2.3 million or more) now own 40% of the country's wealth; the top 20% (whose income is $180,000 or more) own 80% of this nation's wealth. According to the GDP, life has gotten progressively better over the past 50 years. The GPI, however, shows that for the majority of Americans the standard of living rose until the 1970s and has gradually declined ever since. Meanwhile, poverty deepens: in the late 1990s, 21.8% of children live in poverty, up from 16.4% in 1979.[16]

Be the catalyst for your spiritual community putting their beliefs into action.

- *Organize educational programs on the wage gap, wealth distribution, and inequality.*
- *Write a petition to your Congressperson to take action on these issues, and set up a table after services to get congregants' signatures.*
- *Ask your minister to incorporate this into the sermon.*
- *Invite your Representative to speak on what he or she intends to do about the growing wage gap.*

Get talking points from the "Wage Gap Organizing Kit" created by United for a Fair Economy. Send $8 (includes postage) to 37 Temple Place, Boston, MA 02111, or call 617-423-2148.

Rethinking Our Economic Direction

It is now time for us invaders to learn from the native peoples. The Iroquois weighed every decision as to its effect on the next seven generations. If we are to leave a world worth having to our children and grandchildren, we must adopt the long-term strategy of sustainable development.

For 25 years, the economist Herman Daly has been a leader in questioning conventional economic wisdom. Daly differentiates *growth*-the increase by addition of material—from *development,* which is realizing the potential of, or bringing to a fuller, greater, or better state.[17] Our present course, euphemistically called "sustainable growth," is unsustainable: greenhouse gas build-up, ozone layer depletion, and acid rain are but a few of the signs. "Sustainable development"—promoting activities that improve the quality of our lives, such as organic farming, retrofitting homes and offices for energy efficiency, extending rapid transit networks, and building small theaters for community meetings and local drama troupes—is the more viable option.

Heavily taxing resource extraction and energy use would go a long way toward boosting sustainable ventures. Making the two-child family the norm—both in the Third World where families are large and in Western countries where each person's resource consumption is so massive—is a crucial goal.

> *Get people thinking about the implications of growth and the global economy. Download vivid, easy-to-understand posters from Poster Nation at* www.stw.org/posters*and post where appropriate and environmentally desirable.*

Information coming at us from all sides indicates that we need to rethink the very nature of our economic lives. For too long we have operated under these assumptions:

- People are primarily motivated by self-interest and financial gain.
- What's good for big business is good for America. Progress is marked by the accumulation of material goods.
- Competition is the best strategy.

Competition vs. People and the Environment

Is competition really the best strategy? Is it the essential modus operandi for our economic system?

What about the toll it takes on workers and the environment, both in this country and around the world?

The race to be Number One leads U.S. corporations to move tens of thousands of jobs to Third World countries every year. The havoc wrought on American families who depend on this employment is incalculable; divorce, alcoholism, depression, suicide, and child abuse increase dramatically in communities where jobs are lost. Meanwhile, transnational corporations make huge profits in underdeveloped countries where enforcement of environmental safeguards and workplace safety regulations is lax, and wages are often 1/25th of those earned by the lowest-paid U.S. workers.

How much longer can we acquiesce to a system where the CEOs and major stockholders make a killing by moving our living-wage jobs overseas? Who buys all these products made by cheap labor? You can be sure it is not the Third World workers, who earn barely enough to feed their families.

Even those fortunate enough to have a good job feel the impact of competition relentlessly driving globalization. At every wage negotiation, workers are pressed to accept only minor pay increases and take decreases in benefits—under the threat that, if they refuse, the company may decide to move overseas. Many workers no longer have the minimal security of being a company employee. Their status has been changed to independent contractor, and they face a new pressure: constantly having to market themselves. To get jobs at all, they have to bid low. They struggle to make ends meet, with the additional burden of having to pay all their own healthcare costs and file quarterly taxes. Most would like to have their old job back, with its greater stability and benefits, but this is no longer an option.

Workers are not just replaceable parts; they are a principal reason why a company is robust and profitable. The lowering of morale caused by global competition saps both workers' effectiveness and companies' long-term vitality.

We have been told time and again that competition brings out the best in us, but hundreds of studies reveal the opposite. As Alfie Kohn discovered after extensive research, the inventive process is most fruitful in collaboration, where the synergistic interplay of ideas brings creativity to its zenith. Conversely, when we are in the competitive vice grip, we cannot be fully focused on the task at hand, because we are anxiously peering over our shoulder to see who is catching up with us. Razor-sharp competition may be part of the reason, as Dr. Deepak Chopra observes, the majority of heart attacks occurs at 9 on Monday morning.

In the rush to increase sales volume, the corporate push to keep the competitive edge also devastates the environment. The products used and discarded by the average American tap about 500 times more planetary resources than the daily consumption of a poor person in India. If we are not to drive the ecosystem beyond its capacity to regenerate, we must take back the economic sphere—inserting ourselves persistently into decisions and policies about the world of work and money.

Bring stamped, blank postcards with you when you meet your friends for walking or lunch. Take three minutes to write together to your local, state, or national representative about an issue that touches you. Drop them in the first mailbox you see! (See addresses in Resources for Chapter 6.)

Profits vs. the Essence of Life

The very nutritional quality of our food is being compromised by the quest for corporate profit. As Marc Lappe and Britt Bailey discuss in their groundbreaking work, *Against the Grain,* almost half of the soybean crop is now planted with Round-Up Ready Soybeans, produced by the Monsanto Corporation. These have been genetically altered so that the soybean plant is not killed when the toxic herbicide, Round-Up, is sprayed repeatedly on the fields. While the weeds die immediately, the pesticide residue remains on the soybeans, potentially tainting both beef and dairy cattle feed and the soybeans processed for direct human consumption. In addition, as more people eat soy products to increase their phyto-estrogens (proven to protect against osteoporosis, breast and prostate cancer) research on Round-Up Ready Soybeans is showing that this technological innovation has 14% less phyto-estrogens than organic soybeans.[18] This affects even those who have never allowed a piece of tofu to pass their lips, because soybeans are contained in everything from salad dressings and coffee creamers to cosmetics and pharmaceuticals.

In another troubling development, Cornell University entomologists have found that genetically altered corn could well be a culprit in killing Monarch butterfly larvae. The Monarch population is in steep decline due to pesticide use and habitat destruction. In their study, almost half of Monarch butterfly larvae that ate milkweed dusted with genetically altered Bt corn pollen died after 4 days. Bt is a natural bacterium that is lethal to corn-borers and larvae. When its genes are spliced into corn, the corn produces Bt toxins.

Corn is wind pollinated, and its pollen typically drifts 60 to 100 feet in windy conditions. Milkweed, a natural food source for Monarch larvae, often grows near cornfields. About half of North America's eastern Monarch butterfly population is concentrated in the corn belt, and 20% (increasing each year) of the U.S. corn crop is planted in Monsanto's bioengineered corn.[19] Though proponents claim Bt corn reduces pesticide use, organic growers use no toxic pesticides and produce healthy, flavorful corn. Is it really worth endangering the web of life to keep Monsanto's corporate profits high?

Corporate Propaganda in the Schools

Another part of the web of life that is inappropriate for corporate penetration is our children—in the sacrosanct environment of schools. Channel One is a corporation that provides schools with free televisions and cable access on condition that all students watch ten minutes of processed news and two minutes of advertising each day. The televisions are pre-set so that they neither can be turned off nor turned down during the daily broadcast. For advertisers, it is not enough that children are immersed in ads from TV, radio, and billboards from morning until night, seven days a week. To build brand loyalty children must be reached in all their activities. Channel One is no small venture. It is now in 40 percent of all middle and high schools in the United States.

Of course, Channel One is not the only corporate predator. Over 200 corporations shower resource-hungry schools with "instructional" films, textbooks, and computer software. Some of these materials are sent directly to schools, others are given out freely to teachers at conventions. Glossy literature casts global warming as incomplete science and downplays ozone depletion. Posters reframe oil derricks as habitat and clear-cutting as a forest management tool. Consumer's Union, the nonprofit publisher of *Consumer Reports,* analyzed 111 of these corporate handouts: Nearly 80 percent included overt bias, commercial pitches, or inaccuracies. There are also the Coke and Pepsi turf wars over which company will be granted exclusive access to the student body in the guise of raising revenues for the schools.

Is Channel One in your child's classroom? Request that a discussion of the appropriateness of advertising on campus be added to your school board's meeting agenda.

When is enough enough? When will we recognize that we have stepped too far over the line, allowing the profits of transnational corporations to infiltrate every aspect of our lives? Giving our childre the freedom to develop talents, passions, and creativity is not compatible with allowing corporations to amass billions of dollars at their expense. We must fund our schools fully, putting children first, rather than throwing them on the mercy of corporations with a vested interest in exploiting this new generation of consumers.

Is it possible to say no to the explosive push of corporate advertising? All it takes is for us to set our intention. Physicists are now saying what masters of all the greate spiritual traditions have known for centuries: We are all one. We are one interconnected electromagnetic energy flow. Our thoughts and feelings do influence everyone else. Our childrens' learning environment cannot be held hostage to corporate greed.

> *A multitude of informational and training materials on the real impact of corporate penetration are available. Send for "The Growing Divide: Inequality and the Roots of Economic Uncertainty" ($20 including postage) from United for a Fair Economy, 37 Temple Place, Boston, MA 02111, or call 617-423-2148.*

Returning Corporate Charters to their Original Purpose

Corporations are not omnipotent; they are artificial forms that receive their life from a state charter. Of course, the average educated person has no inkling of this. Richard Grossman and Frank Adams, in their booklet *Taking Care of Business: Citizenship and the Charter of Incorporation,* thoroughly detail how corporations have gradually taken power and what we can now realistically do to stop them.

In the early days when states issued charters, corporations were obliged to obey all laws, serve the common good, and cause no harm. The charters were granted for a limited number of years. Corporate managers and stockholders were liable for harms, injuries, and debt, and charters could be revoked for infractions. Unless the legislature renewed an expiring charter, the corporation was dissolved and the assets were divided among the stockholders. Interlocking directorates were outlawed, and corporations were prohibited from owning other corporations.

Big changes came in the years following the Civil War. Corporations availed themselves of a little help from their friends, in this case politically appointed judges. These judges freely reinterpreted the U.S. Constitution and transformed common law doctrine.

They eliminated jury trials whose purpose was to determine whether corporations caused harm and to assess damages. They ruled that workers were responsible for their own injuries; this became known as the "assumption of risk." They gave corporations legal justification to deny First Amendment rights of free speech and assembly on company property.

The most damaging interpretation came in 1886, in *Santa Clara County vs. Southern Pacific Railroad.* The Supreme Court ruled that a private corporation was a "natural person" under the U.S. Constitution, and thus was sheltered by the Bill of Rights and the Fourteenth Amendment, which was initially adopted to protect freed slaves. Using the Fourteenth Amendment, the justices struck down hundreds more local, state, and federal laws that had enforced some degree of corporate accountability to citizens. In the ensuing years, judges even prohibited unions from striking.

While corporate power today seems monolithic, a thoughtful look can give us great hope: the basis of the corporate structure is still the state charter. The interpretations that have come down through the years have all built upon the rulings of judges who were appointed at a time of flagrant political corruption growing out of the power of the robber barons of the railroad, mining, and newspaper industries. Grossman and Adams remind us that we can stop limiting our thinking only to existing law, stop accepting corporate definitions of property, and reject laws made by fallible judges that grant corporations more rights than humans and allow the destruction of the ecosystem upon which we depend for survival.

Democracy—Use It or Lose It!

We are, after all, a democracy! We can make laws that serve the common good. We are simply out of the habit of being active citizens. Grossman and Adams suggest a range of responses we can make:

- Ban corporate donations in elections.
- Buy stock. Stockholders have the authority to seek injunctions and file corporate dissolution suits if managers are acting illegally, oppressively, or fraudulently, or misusing corporate assets.

- Exercise our right as citizens to amend and revoke the charter of harm-causing corporations.
- Give workers free speech and other Bill of Rights protections.
- Once again make managers, directors, and shareholders liable for corporate debts and harms.
- Prohibit corporations from buying and owning other corporations.
- Change judge-made laws such as doctrines that declare corporations to be legal persons.
- Place ceilings on the tax-deductibility of excessive corporate salaries.
- Require increasing levels of worker and community ownership.
- Ban corporations from advertising on public policy and legal issues.
- Get corporations (such as Channel One) out of our schools.
- Forbid corporations from hiring permanent replacement workers in strike disputes.
- Ban the use of deadly chemicals in production processes.
- End tax deductions for legal fees, advertising, and fines.
- Increase corporate tax rates.
- Redefine workers' rights and community rights.
- Bestow broad legal standing on species and the natural world.[20]

William Greider, author of *Who Will Tell the People?,* adds three more suggestions:

- Institute a system of graduated penalties for corporations who repeatedly pollute the environment or defraud the government.
- When Congress enacts tax benefits, stipulate that no corporation is eligible if it has violated laws and regulations in the preceding years.
- Withhold tax credits for research and development from corporations who shrink their American workforce.

Looking at the big picture of our corporate-dominated world can seem overwhelming. Picking any one of these logical steps, and connecting with others already involved, makes change seem more possible (see the back of the book). As a matter of fact, you can make a big difference starting right now: simply buy locally. For example, buy books at your local bookstore instead of the giants that are putting the small, independent booksellers out of business.

You may pay a little more to keep local merchants in business, but the money keeps circulating in your own community rather than being sent thousands of miles to corporate headquarters to go into CEO mega-salaries and shareholder dividends. Get your cappuccino from the neighborhood cafe rather than Starbucks. Buy your art supplies from the local art store that provides Saturday classes for kids rather than driving those extra miles to the discount supplier. Buy that new tool from the mom-and-pop hardware store rather than Wal-Mart. Purchase your next computer printer from the nearby shop run by computer nerds—when you need help or servicing, they're *real people* to talk to.

Gandhi on Self-Sufficiency

Mahatma Gandhi was perhaps the most inspiring individual of the twentieth century. The insights of his brilliant mind and great heart are still relevant to our dilemmas at the turn of the millenium. But he was not always clear and powerful. In his early life he had a raging temper to contend with and, as a lawyer, he so bungled his first court case that he ran out of the chambers, humiliated and tonguetied.[21]

Most of us connect nonviolence with Gandhi, but the term has been distorted. "Nonviolence" does not mean passivity; rather, it is active engagement using no harmful means. Satish Kumar describes an even more central operating principle for Gandhi, *swadeshi,* or local self-sufficiency:

> For Gandhi a machine civilization was no civilization. A society in which workers had to labor at a conveyor belt, in which animals were treated cruelly in factory farms, and in which economic activity necessarily led to ecological devastation could not be conceived of as a civilization. Its citizens could only end up as neurotics.[22]

Gandhi's focus was not individual gratification but the longterm welfare of the community. He acknowledged that a certain degree of physical comfort was necessary but felt that material objects too easily became a hindrance rather than a help. He felt that work by hand brought a meditative mind and genuine self-fulfillment.

Gandhi saw that "the masters of the money economy want more and more efficient machines working faster and faster, with the result (being) men and women thrown on the scrap heap of unemployment."

Speaking the Unspeakable

Gandhi's timeless words describe the state we find ourselves in today, dominated by bottom-line-driven multinational corporations. New technologies eliminate more jobs than they create. The earth is reaching its capacity for being able to absorb the toxic wastes that industry and motor vehicles expel into the air, the water, and the soil. Debilitating health problems are on the rise: cancers, immune deficiency diseases, and brain tumors. Jacques Ellul noted that, for every problem science and technology seem to solve, ten unforeseen consequences arise, many of which are worse than the original malady. DDT, for example, started out as the wonder pesticide, then drove many species to the brink of extinction, and now has dispersed so widely that it can be found in penguins in Antarctica.

Corporations are driven by profit and growth imperatives. While their beguiling ads promise us ease, comfort, and excitement, for most people the

American Dream is receding further into the distance as they work longer days in an ever more frenzied time frame, have fewer hours for themselves and their loved ones, and watch the planet become more fouled by pollution.

Perhaps we can let Gandhi's struggles and courage infuse and fortify us so we can call multimillion-dollar CEO compensation packages by their real name: *greed.*

The Interfaith Center on Corporate Responsibility publishes a monthly newsletter that summarizes various shareholder resolutions and tracks the vote on them at more than 100 public companies. Call 212-870-2293 to see which corporation you'd like to buy stock in, thereby being able to vote directly on socially responsible resolutions. Propose this to your spiritual community, support group, family, or friends.

Perhaps we can find our voices and begin to talk about where the record corporate profits are coming from: employees' decreasing wages, healthcare, and pension benefits. Perhaps we can speak the unspeakable, and bring economic justice into conversations around the office water cooler, into our national debates, and into our electoral campaigns.

Questions For Reflection

- How do you react to: "In reality Mr. Sam was in the business of driving small shops out of business and paying most of his workers minimum wage, while he amassed a personal fortune of $23.5 billion.... Most employees are given only part-time work so that the company can avoid paying the benefits fulltime workers must receive"?
- Would you be willing to pay a little more for products and services to keep a local merchant in business rather than going first to a big chain store?
- Do you think pollution, crime, and ecological depletion should be reflected in our national economic indicator?
- Gandhi felt that a society in which economic production and profit was the highest priority created neurotic people. To what degree do you think this is true of America today?

Chapter 3

The Hidden Costs of Competition

Suzanne Stoddard

the heavy price we pay to win

Winning isn't everything; it's the only thing.
—Vince Lombardi

You're not winning the silver, you're losing the gold.
—Nike Corporation

To conquer oneself is a greater task than conquering others.
—Buddha

The joy is in the doing.
—unknown

If something is worth doing, it is worth doing poorly.
—Benjamin Franklin

Competition is as American as hot dogs and football. It weaves itself subtly into nearly every aspect of life in the United States, structuring not only work and sports, but also our social lives, our family time, and often our recreation and vacation hours. Competition is so sacred to our way of life that, most of the time, we don't even think to question it. Instead, we submit to the punishing stress it creates. Of course, we all know that Little League can get ugly—we've seen coaches and parents yelling with frightening ferocity at kids and umpires. We've all, no doubt, been critical of moms and dads who seem to be living their lives through their children, hoping to make up for their own unrealized dreams. But we shrug these off as extreme cases, as the price we have to pay for our biology and for living in the kind of world we do.

The Competitive Fabric of Our Lives

Research by anthropologists across a wide spectrum of cultures, however, shows that competition is learned rather than innate. The seminal collection of anthropological research, *Cooperation and Competition Among Primitive Peoples,* edited by Margaret Mead and originally published in 1937, concluded that the use of a competitive ethic varies widely across cultures and is culturally rather than biologically determined.

The socialization process that indoctrinates youngsters begins in earliest childhood and is usually very successful in conveying and reinforcing the belief system of the group. Research since Mead's time has bolstered this conclusion. For example, the Inuit of Canada have almost no competitive interactions, with work and play being almost entirely cooperative. Australian aborigines have been discovered to cooperate as willingly with members of other tribes as with those of their own tribe.[23] In another study, Israeli children on a kibbutz (communal living situation) found cooperation more effective than competition, while urban Israeli children persisted in competition even when there was no positive result from it.[24]

Enjoy a noncompetitive board game: Buy "The Ungame" or "Life Stories." Every turn provides another opening into hearts and minds, with questions like: What do you think is your greatest personal achievement to date? What do you value most in life? What is one thing you would like to improve about yourself? What do friends like about you?

While competition has probably always existed in some form, somewhere in the world, the modern corporation has perfected it. Alfie Kohn, who has spent the last eighteen years researching competition, has said about American culture: "Our collective creativity seems to be tied up in devising new ways to produce winners and losers.... The only way we can think of to socialize with the people who work for another company is to try to beat them in a competitive game.... No corner of our lives is too trivial—or too important—to be exempted from the compulsion to rank ourselves against one another.... Our lives are not merely affected by, but structured upon, the need to be 'better than.'"[25]

For one day, pay special attention to competitive tendencies that have become ingrained in your behavior, actions, or words, at work and at home. Ask yourself if they are serving you well.

Where did we get the idea that we have to beat other people in order to be our happiest, most actualized selves? When the ancient Greeks developed the Olympic Games, the point was for the competitors to achieve their personal best, not for each to measure himself against the others. Might it be possible to go back to this concept—to learn self-discipline, produce excellent results, and have fun—without making those around us lose? Before attempting an answer to this, let's take a closer look at what competition is doing to us.

When Sports Are Not Play

Today our children's play lives are dominated by scheduled games, meets, and practices. Many parents are asking, how much is enough? When is the pressure too much? Should a child really be urged to "play through injuries"? And how often do children participate in a competitive sport because they think their parents expect them to? Probably a lot more than we realize. A growing number of parents are ambivalent about their kids' involvement. Although they see some real benefits for their children in the discipline and athleticism developed in learning to play a sport at a high level, many admit that organized sports take up too much family time, and they're secretly pleased when a game is rained out or their child's team doesn't make it to the championships.

One father I know says his sons from a previous marriage don't know how to "just have fun." Every recreational activity is seen as an opportunity to beat someone. When he takes the boys to a swimming pool, all they want to do is race him or each other to the finish. Once, when he took them to a beautiful mountain lake on vacation, they were at a total loss as to how to amuse themselves because there were no lanes to swim in and there was no way they could invent to beat each other. David Elkind, author of *The Hurried Child,* says that, "unfortunately, both the value and the meaning of play are poorly understood in our hurried society. Indeed, what happened to adults ... has now happened to children—play has been transformed into work. What was once recreational—sports, summer camp, musical training—is now professionalized and competitive in schools."[26]

Do something you enjoy but normally don't do because you're not "competitive" at it. Feel the liberation!

Too many children have negative experiences with competitive sports and games at school, because they always finish near the bottom in timed athletic events and are the last to get picked for teams. Some develop complexes that last a lifetime. By the time a young person is in high school, if he's not good enough to be on the competitive team, there aren't any other options for playing that sport just to stay in shape and have fun.

The story of Ellen's son and the Pine Box Derby is a poignant, all-too-familiar example of competition teaching the wrong lessons to both winners and losers. Damian, at age 8, was eager to build a model car for the Boy Scout contest. All the boys were given a wooden block out of which they were to carve their car with only a minimum of family help. When the evening of competition came, Ellen, her husband, Michael, and Damian, proudly clutching the car he'd worked so hard on, arrived at the school auditorium in excited anticipation. All contestants took their cars to the top of the slope and released them when the bell sounded.

A few of the kids had brought sleek, highly engineered cars that had obviously been made by adults using precision tools. These cars, of course, whizzed in a straight line down the track. The rest of the cars, crafted by the children themselves, as they were supposed to be, knocked back and forth between the side rails, rolled unevenly down the track, and came in last. The high-tech cars handily won the prizes and the winning boys exulted in their victory. Damian and the other youngsters, who had begun the evening feeling pride in the cars they'd made, left the auditorium blinking back tears. Two painful lessons had been learned: anything I make myself is no good, and winning is more important than following the rules.

One hurtful incident of "losing" does not a complex make. But so often the labels children perceive others putting on them, or those they place on themselves early in life, get reinforced in every new situation structured as win-lose. For boys, the playground and sports arena are most often the crucible where stigmatizing labels are internalized. A very insightful analysis I read in the newspaper after the Littleton, Colorado high school massacre discussed the Second Amendment's "right to bear arms" in relation to those branded by society as losers. The author, Bob Wieder, noted that these "Last Laughers" often carry very deep grudges against "winners." In an era when guns are accessible even to teenagers, obsession with revenge can cut a wide swath of destruction.[27] It is telling that the killers, who had previously professed hatred for "jocks," targeted a coach and several athletes, among others.

It is so often said that competition builds character that we seem to believe it reflexively. However, psychologist David Walsh, author of *Selling Out America's Children,* tells a story where the need to win became so all-consuming to a Midwest youth hockey team that a young girl tragically died. The team had traveled out of town for a game. In the hotel where the players spent the night, an eighth grade cheerleader was raped by several team members. When she reported the crime, the girl was ridiculed and harassed. She had done the unforgivable—made accusations against star players, without whom the team couldn't win. Many adults in the community blamed the girl herself for the rape and criticized her for pressing charges. She was accused of damaging the future careers of the young men. Three years later the girl committed suicide, having been continually persecuted by those for whom the truth was too big a threat.[28]

Jeff Benedict and Don Yaeger reveal that breaking the law and perpetrating violence are no impediment to premier athletes' having a professional sports career. In their 1998 book, *Pros and Cons: The Criminals Who Play in the NFL,* Benedict and Yaeger show that one-fifth of National Football League players have arrest records for crimes ranging from drug charges to domestic violence to rape and even murder.[29] Violence has become so accepted in American life that even criminals are privileged to play on prestigious teams and command dream salaries, because entertaining the public is such big business.

Where do you see violence accepted and even glorified? Write a Letter to the Editor of your local paper saying how you feel about this.

World Class Losers

Every two years, much of our national attention is riveted on the Olympics. In the summer of 1996, a particular Nike commercial stunned me. In this television ad, the second-place finisher of an Olympic event was told: "You're not winning the silver, you're losing the gold." It was irrelevant to the advertiser that these "losers" were the acknowledged second best in the *world* in their event, and that they had spent many years of their lives in pursuit of peak performance through grueling daily workouts, Zen-like focus, and enormous personal sacrifice. The ad's blatant message was chilling.

What do these words and images teach our children? Is it a value system we want them to absorb?

The Tonya Harding – Nancy Kerrigan incident from the 1994 Winter Olympics is an extreme example of the lengths to which competitors have gone to be Number One. Thus, we may view the appalling attempt by Harding's boyfriend to disable Kerrigan just days before the ice-skating competition as an isolated example of competition gone completely awry; yet the seeds for this kind of behavior are planted every day in the barrage of messages glorifying winners—from television programs, advertisements, and the mouths of parents and coaches.

Next time you watch television commercials, notice how personal worth is equated with wealth, height, slimness, blondeness, speed, the kind of car a person drives, or being Number One in something.

George Foreman, former world heavyweight boxing champion, said in his autobiography that winning the boxing title had made him obsessed with staying the best in the world. It poisoned all of his relationships, with women as well as men. "When you're Number One, everyone wants to challenge you, and the pressures to keep your title are unbelievable. All I was was the best. I couldn't go higher. The only direction was down. All I could do was keep trying to defend my title." When he was ultimately defeated by Muhammed Ali, his life fell apart. He felt he'd been stripped of his manhood, because his entire sense of self-worth came from being the best heavyweight boxer.

Although there are many professional athletes who are terrific role models for children, giving their time to boys' clubs or helping coach inner-city youth teams, an equally large number send kids the message that money is everything when they desert their local team to pick up a higher salary with another club. Such athletes often sneer at the idea that they should feel loyal to the community that supported them. Professional sports is, after all, big business. The most talented athletes charge whatever they can get away with, knowing that savvy team owners will pay top dollar; the amount of money the players can bring in at the box office makes it all worthwhile.

The degree of salary escalation seen in the professional sports world in the last several years is as absurd as that in other areas of the entertainment industry. Why is it that investors can nearly always put together a deal with city governments to finance a multimillion dollar stadium, but schoolchildren have to hold bake sales to raise $300 for after-school sports equipment and jerseys?

The Very Big Business of Pro Sports

The economic stakes in today's spectator sports are mind-boggling. The competition to be the host site for the 2002 Winter Olympic Games is a case in point. Salt Lake City officials not only wined and dined International Olympic Committee members, but offered complimentary four-year tuition scholarships at Utah colleges to members' children. Once these unsavory bribes hit the headlines, a few resignations ensued, but Salt Lake City was allowed to remain the host site, guaranteeing lucrative revenues for the city and its businesses. The message to our children is unambiguous: cheating pays.

Clearly, the chief beneficiaries of the American mania for competitive sports are not the spectators watching TV from their sofas, but those with a piece of the financial action. During the 1999 Super Bowl, advertisers paid a record $50,000 *per second* of air time. Increasingly sophisticated commercials plant subtle messages in viewers' minds, and although most people insist they pay no attention to advertisements, Americans carry the greatest amount of personal debt in the world.

In the documentary film *Manufacturing Consent,* Noam Chomsky decries our national obsession with spectator sports and passive television-watching as a distraction that keeps people from taking action together to improve the quality of life within our communities and nation.[30]

> *Instead of spending two hours watching a sport on television, volunteer to coach kids who want to play an organized sport but don't qualify for the team at school.*

Myths We Love to Believe

While most people believe that competition is destructive when carried to extremes, few would probably assert that competition is negative in itself.

However, in *No Contest: The Case Against Competition,* Alfie Kohn convincingly refutes every tenet our society has internalized about the benefits of competition and concludes that:

- competition is *not* an integral part of human nature.[31]
- competition does *not* bring forth greater excellence than cooperation.[32]
- competition is *not* more enjoyable than cooperation.[33]
- competition does *not* build character.[34]

Kohn says that he began his examination of competition believing that some competition was good, but after reviewing over 2,000 studies found that cooperation produced far more benefits and higher levels of achievement than competition. This was true whether the arena was athletics, medical research, education, politics, education, or even business.[35] Researchers, inventors, think tanks, and political parties waste enormous amounts of collective energy and time when they act in competition. It is the synergy of collaboration that raises human achievements and solutions to the highest levels. Stephen R. Covey has written at length about this in *The 7 Habits of Highly Effective People.* Covey says:

> When properly understood.... synergy is the highest activity in all life.... The highest forms of synergy focus.... the motive of Win/Win, and the skills of empathic communication on the toughest challenges we face in life. What results is almost

> miraculous. We create new alternatives—something that wasn't there before.... Synergy is everywhere in nature. If you plant two plants close together, the roots commingle and improve the quality of the soil so that both plants will grow better than if they were separated. If you put two pieces of wood together, they will hold much more than the total of the weight held by each separately. The whole is greater than the sum of its parts. One plus one equals three or more. Synergy.... catalyzes, unifies, and unleashes the greatest powers within people.[36]

There is no basis in reality for the idea that many must lose so that others can win. It is time to let go of worn-out ideas that were never true about how people are and how the world works.

Old Wisdom, New Directions

Competition is often considered a positive attribute of a culture in which the individual is highly valued. Paradoxically, however, competition leads to a high degree of conformity. Individuals who don't enjoy competitive games are often branded as wimps who "can't take the pressure." Maybe we should admire such people instead. Thoreau said "If a man does not keep pace with his companions, perhaps it is because he hears a different drummer. Let him step to the music which he hears, however measured or far away."

If Thoreau hadn't followed his own advice and lived in the noncompetitive way that was right for him, we surely wouldn't have *Walden,* a classic of American literature, full of timeless wisdom for simplifying one's life. Nor would we have his elegant *Civil Disobedience,* one of the main sources for Gandhi's campaign of nonviolence, which in turn inspired Martin Luther King and the civil rights movement.

Many of us love adrenaline, and we may be convinced that competition is what keeps us honed and toned. But the evidence is in, both from researchers and from the stress-related symptoms showing up in Americans at ever younger ages: competition causes more problems than it solves.

It is never too late—indeed it is always the right time—to find a better path. As the Chinese proverb says, "If you don't change your direction, you're likely to end up where you're headed." There are much better games out there. Let's discover them, play them, and live them.

Questions For Reflection

- In what ways are you competitive? Where does competition take a toll on your body?
- What does competition do for you that's positive? In what ways does it detract from your life? How can you eliminate the negative competition?
- When you are competitive, is it about things that really matter? Is it when driving, at work, in sports

or games? Can you imagine getting the same rush of adrenaline in a healthier way?

- Can you envision your life without competition? What would be different? How would it feel?
- What if you and your family decided to step outside the rat race? What are the activities or product purchases you could eliminate?
- Remember a time when you felt like a loser or made someone else feel that way. What were the circumstances? Does any good come from feeling worthless or from cutting someone down to size?
- Does television foster the win/lose belief system?

Chapter 4

This Is Entertainment?

Ellen Schwartz

TV as purveyor of a culture of disrespect and promoter of a passive populace

All television is educational. The question is, what does it teach?
—Nicholas Johnson, former Federal Communications Commissioner

Violence grabs the headlines, but violence itself is the result of a society that promotes selfishness, greed, and instant gratification.
—David Walsh, *Selling Out America's Children*

Television is the most pervasive element in our lives as we near the end of the second millenium. Rich or poor, we all love our TVs. Sixty-six percent of homes have three or more televisions. As Americans, we feel it is our birthright to be able to enjoy the "free" entertainment of television. But what is presented as entertainment is overwhelmingly graphic violence, flagrant disrespect, casual sex, and unending messages to buy, buy, buy. The price may be free, but the cost is high.

The High Cost of "Free" TV

Television has changed enormously in the past three decades. Any flip of the remote control will serve up countless images of violence. It used to be that you could tell the difference between the good guys and the bad guys. Now the good guys use violence almost as frequently as the bad to achieve their ends. By the time our children have graduated from high school, they've seen 200,000 acts of violence on TV.[37] What makes this situation all the more dangerous is that over 70% of the violence is shown with *no negative outcomes.*[38] Kids do not see the realistic consequences of debilitating injuries, shattered lives, and broken homes and families. Even the substance of comedy shows is put-downs. Kids don't watch TV and think, "This isn't how you talk in real life. This isn't how you should treat people." They just absorb: *This is the way the world is. This is normal.*

What does it mean for our children to be inundated by these kinds of stories? From time immemorial, we have learned through stories about our culture and about each other—what to do, what not to do, and what the consequences of our actions may be. Television, the main purveyor of stories today, supplies a different kind of message, as well as violent details for re-enactment. Teachers doing lunchtime playground duty often see repeats of last night's action shows, with their kicks, punches, and put-down language.

Sure, most kids don't grow up to be assault-weapon murderers, but the effects of TV are real, impacting how they relate to others and how they solve problems. I remember speaking with a group of junior high and high school teachers. One teacher said: "Do you know the difference between kids now and fifteen years ago? Kids now do very hurtful things to each other *with no feeling.*" The other teachers nodded in agreement.

Psychologist David Grossman, a retired Army officer, thinks that point-and-shoot video games may be murder simulators for young people. During World War II, the Army started using man-shaped targets to break down recruits' aversion to killing. Video games give extra points for head shots. Paducah, Kentucky schoolboy killer Michael Carneal coolly shot nine times, hitting 5 of 8 victims in the head or neck. Throw away any violent video games you find in your house.

Do we really think that kids who watch and listen to hour after hour of cutting remarks, threats, punches, kicks, and mutilations will automatically know how to work out problems with real people? How to listen to someone whose point of view is different from theirs? How to tune into someone else's feelings? How to be compassionate to what others have to deal with and try to help them?

Children need to see realistic models of conflict resolution again and again, so they absorb it as second nature. They need to see the process of: letting each person present his or her view of what happened; checking in on everyone's feelings; discovering what intentions were; exploring "this reminds me of...." and "this always happens to me when..."; finding common ground; brainstorming how the situation can be win-win; and then picking a solution agreeable to everyone.

Over 2,000 studies have been conducted on the impacts on children of watching violent television programs. The solid majority show a strong correlation with increased disrespectful behavior, verbal abuse, and physical aggression. Even the cable TV industry corroborates these findings! In 1996, the industry released the results of a $1.5-million dollar study that examined 2,500 hours of programming. They found that violence is pervasive across the spectrum of both network and cable offerings, and that it is psychologically damaging to children. But did they actually reduce the violent programming? Not a chance!

> *Find out if your local primary or secondary school has a conflict resolution program. If so, and if you have a child, encourage him or her to be a peer facilitator. If not, would you be interested in working with others to get one started?*

Remember the age-old adage: follow the money? The majority of entertainment industry profits come from overseas sales. Relationships and humor, which make a story really interesting, are very culture-specific. Violence needs little translation, however. Anyone can understand someone being hurled through the window of a high-rise building.

What Kids Are Getting and What They Are Not Getting

There is now an epidemic of sleep disorders among children. Frightening images seen on television or at the movies flash graphically into memory when it's late at night and the house starts to creak. The movie *Jurassic Park* had many terrifying scenes. Nonetheless, the marketplace was flooded with companion items for children: sheets, undies, lunch boxes, T-shirts.

When Terry Gross was interviewing the director, Steven Spielberg, on her PBS radio program *Fresh Air,* she asked how his children reacted to the movie. When he responded that he wouldn't let them see it, Gross was speechless. It's okay for Spielberg to make millions of dollars off of *your* children's terror, but his own children, he protects.

In *Four Arguments for the Elimination of Television,* Jerry Mander talks about another troubling aspect. Seminal research seems to be showing a similarity between the electromagnetic color frequencies emitted by televisions and those found in the artificial colors and flavorings in food. Strong correlations have been observed in many children between eating such foods and hyperactivity.[39] Parents frequently notice that their children are uncooperative, rowdy, or aggressive after they have been watching television.

While watching violent images is damaging on many levels, even worse is what the children are missing out on while they're sitting in front of the tube 24 to 28 hours a week. Preschool teachers will tell you that the youngest ones are not getting their developmental needs met. Small-muscle coordination is developed by grasping a crayon and making a big red "A" on the paper, not by watching the letter "A" dance across the screen. Large-muscle coordination comes from riding bikes and climbing trees, not from watching junk food commercials where other kids are playing and running.

For this and numerous other reasons, the American Academy of Pediatricians now recommends that children watch no more than one or two hours of TV or videos per day. The impacts are potentially very serious, especially for preschoolers. If a growing brain does not receive the appropriate stimuli in the correct time frame, the neurological pathways do not develop and it is much more difficult to learn certain things later in life. Television uses very simple sentence structures and a limited vocabulary. Real-life conversations that a child could hear, or join, are much more complex. Children watching too much television are in danger of receiving insufficient stimuli for full development of their linguistic pathways.

TV Watchers Are Harder to Teach

Teachers are in a particularly difficult situation in the Age of Television. On TV, the camera angle (or entire scene) changes every 7 to 10 seconds. MTV's images change even more frequently. Whether working at the preschool or high school level, the teachers' main challenge is keeping kids' attention. Many children are bored in school because they have become attuned to this rapid visual change, with which no educator can compete. Real learning is accomplished at the much slower pace of natural rhythms. Children are often frustrated when they can't solve a problem or write a paper instantaneously; they are shocked when they can't master a musical instrument in just a few weeks.

As founder and executive director of *Healing our Nation from Violence,* a nonprofit entity devoted to educating the public about the entertainment industry's real-life impacts on youth behavior, I recently spoke at an exclusive private school in a San Francisco Bay Area suburb. Afterwards, one of the eighth grade teachers came up to me. She told me that she goes to great lengths, when introducing a new topic, to approach it from many different angles and address the different learning styles of her students. After the fifteen or twenty minute overview, she invites the students' reactions. Two decades ago, she said that students' hands would shoot up right away.

In contrast, today's students only ask questions or make comments after a great deal of coaxing. This teacher and others said that our youths' ability to analyze concepts and relate ideas one to another has been impaired by spending the predominance of their nonschool hours passively watching TV. College professors tell me that their students can bring forth a myriad of *facts* from Internet research, but they are weak in exploring the *connections* between various pieces of information and drawing logical conclusions from data.

Find out if your local school participates in National TV Turn-Off Week in April. If it does, call or write a short letter of appreciation to the principal. If not, would you consider initiating it? TV-Free America *has easy-to-use organizer packets for $10. Call 202-887-0436.*

Our Changing Heroes and Values

In *Amusing Ourselves to Death,* Neil Postman writes at length about "the decline of the Age of Typography and the ascendancy of the Age of Television." He shows how dramatically the content and meaning of our lives has changed "since two media so vastly different cannot accommodate the same ideas."[40] We are now immersed in flashy images and triviality. In the past, children's heroes were Abraham Lincoln, Eleanor Roosevelt, or Martin Luther King, Jr.. Today's children are more likely to imitate Madonna or Michael Jordan. Why? Because technologically beguiling images of movie stars and sports figures flood their consciousness from all sides: TV, movies, magazines, and billboards.

There is a negative symbiosis between today's heroes and mere mortals. Ordinary people, by and large, feel they can't possibly paint their lives on such a big canvas; most give up trying to develop their own creative talents and settle for buying the CDs and videos of the stars. Ironically, those we exalt are also hurt by hero worship. Under enormous pressure to keep up the illusions their public holds about them, they often lose touch with who they are.

In his exquisitely insightful book, *Selling Out America's Children,* David Walsh points out that the values promulgated on commercial TV are strikingly different from what our children need to grow into happy, healthy adults. Television promotes instant gratification, me first, win at any cost, get all you can for yourself, wealth equals happiness, and violence is amusing. The values parents want children to internalize in order to become caring, competent, contributing people are more along the lines of self-discipline, delayed gratification, respect for self and others, generosity, fairness, cooperation, peaceful conflict resolution, and the knowledge that happiness comes from within.[41]

Not only do the major networks present a distorted value system to our young, minority children see almost no one of their own race in roles representing wealth, education, or achievement. More often, people of color appear as criminals or drug addicts, impoverished or acting foolishly. When the four major networks announced their 1999–2000 schedule, not even one nonwhite leading character or theme could be found on any of the 26 new prime time television series.

Write a letter praising a quality program or criticizing something offensive or inappropriate. Consult the TV network address guide in the resource section at the end of the book.

Whose Freedom Is Impinged?

Television and movies do not reflect reality, they create it by magnifying callous and disrespectful behavior, fostering passivity, and increasing the fear factor in our lives. It is the people who have the most to lose financially (network executives, actors, writers) who yell "free speech" the loudest when congressional inquiries or newspaper editorials shine the spotlight on television's contribution to immorality and amorality. Commercial speech, however, is in a very different category from individual speech rights. As an individual, you can't go into a crowded theater and yell "Fire!" but we have allowed the entertainment industry to present whatever they want on the *public* airwaves to increase their profits. The low-grade fare they present is all the more insulting when you consider that their broadcast spectrum, from which they make millions of dollars each year in profit, is leased to them *free* "in the public interest."

"Okay," you say, "I don't watch that junk, but I have to watch the 10 O'Clock News." Yet the line between programs and news has become blurred. For more than half of the news hour, words and images about rape, robbery, murder, war, and scandal assault us. When our "window to the world" is filled with stories that invoke fear and anger, it's easy to feel hopeless and helpless to do anything.

While the major content of the news is troubling, what is omitted may be even more important. We have grown up believing that the press is the fierce watchdog of the public interest. Yet information that the public needs to know to make informed decisions about major problems and issues confronting us is not being presented on the nightly news—the place most people go to keep current with local and global events.

To reach a mass market, great sums of money are needed for both exposure and distribution. As Ben Bagdikian notes, "Modern technology and economics have quietly created a new kind of central authority over information."[42] Conflicts of interest between the public's need to know and the corporate desire for a positive image have vastly increased.

The boards of directors of print and broadcast news organizations include executives from Ford, General Motors, General Electric, Exxon, Alcoa, Coca Cola, Philip Morris, ITT, IBM, and the like. The prime stockholder for the country's major wire service, Associated Press, is Merrill Lynch. NBC is owned by General Electric, CBS is owned by Westinghouse, Ted Turner's CNN was taken over by Time Warner, and Rupert Murdoch controls Fox.

In 1995, Disney bought ABC and immediately forced that network to recant its exposé of cigarette manufacturers deliberately manipulating nicotine levels to increase addiction. Part of the equation was Philip Morris' threat to withdraw its $100 million a year in advertising. Though cigarettes can't be advertised, they own Kraft Foods and Miller Brewing.[43]

How much influence do the owners and chairs exert upon operations? In an interview, Rupert Murdoch was asked, "You're considered to be politically conservative—to what extent do you influence the editorial posture of your newspapers?" He responded, "Considerably.... My editors have input, but I make the final decisions."[44]

During the days when Watergate was beginning to break, a phone call from the Nixon White House was all it took to convince CBS chair William Paley to scale back Walter Cronkite's plan to do an extraordinary two-part series on the break-in before the election. Did this omission make a difference?[45] Nixon was re-elected by an overwhelming margin and wasn't forced to resign until two years after the burglary.

So much has shifted in the delivery of our news that we cannot even be sure whether what we are seeing is genuine reporting or the product of a public relations firm. As a matter of fact, *about 40% of all news flows virtually unedited from public relations offices.*[46] Public relations firms operate largely beyond public view. The 150,000 public relations practitioners in the United States outnumber the country's 130,000 reporters,[47] and with media corporations downsizing their newsrooms, the gap is widening. In *Toxic Sludge Is Good for You,* John Stauber and Sheldon Rampton point out that the activities of today's PR firms go well beyond catchy slogans and clever delivery. They create national and global campaigns using strategic blends of paid media (advertising) and "free media" (public relations). They use 800 numbers and telemarketing, advanced databases, computer bulletin boards, simultaneous multi-location fax transmissions, and "video news releases."[48]

> Video news releases are entire stories which are written, filmed, and produced by PR firms and are transmitted by satellite feed to hundreds of television stations around the world. They are designed to be indistinguishable from real news, and are typically used as "story segments" on TV news shows without any attribution or disclaimer indicating that they are, in fact, paid subtle advertisements. "Most of what you read on the paper and see on TV is, in effect, a canned PR product. Most of what you read in the paper and see on television is not news, "says a senior vice president with Gray and Company public relations.[49]

Mark Dowie, former editor and publisher of *Mother Jones* and the recipient of fourteen major journalism awards, says "The methods of modern advertising, steeped in subliminal psychology and imagery and aimed at the subconscious ... are used to promote and protect ideas, policies, candidates, and hazardous products.... The consequences for our culture, democracy, and public health are staggering to contemplate."[50]

The next time you're watching a TV newscast, let your mind start churning out questions. Why don't we get timely stories about the *hundreds* of accidental radiation releases by our nuclear facilities? Could it be because General Electric and Westinghouse were involved in nuclear production, and the communities around their facilities are still suffering from radiation-induced illnesses (one-millionth of a gram of plutonium can cause lung cancer)? Why don't we hear about the devastation all-terrain vehicles (ATVs) are wreaking on our public lands? Not only do they cause erosion, stream sedimentation, destruction of endangered plant species, and wildlife disturbance, but in Yellowstone Park on a winter's day snowmobiles spew out more air pollution than all automobiles driving through the park the entire summer. Annual sales of ATVs have tripled since 1991.[51] Perhaps the fact that their manufacturers are major network advertisers has something to do with the lack of coverage.

Why aren't there more news stories presenting carpal tunnel syndrome as the most frequent serious injury for workers? Perhaps because it would compromise the profits of their major corporate advertisers to re-outfit their offices with ergonomic furniture and reorganize the work day to include a more interesting (and healthy) variety of activities. Why is there no regular labor anchor on the news hour to balance the Wall Street profits focus? Perhaps because so many corporate sponsors have spent millions driving unions out of existence.

How would our lives be different if the network news hour presented a full array of information—not just the stories that don't offend their advertisers? What if, instead of taking so much time to cover far-away wars and scandals over which we have little or no control, they featured more local problems in which people watching could become involved? Seeing an in-depth story about how a community became concerned enough over toxic emissions to educate themselves and clean up their air could lead viewers to do the same where they live. Or what about watching on TV the magic created when a preschool is built next to senior housing, with elders reading stories and sharing giggles with the children? This could be replicated in almost any community! Television has an enormous potential to create positive ripple effects instead of the mostly negative ones being disseminated today.

Create your own positive news story! Highlight something you're involved in within your local school or community. Call a reporter whose previous stories are in alignment with yours. Think of dramatic visuals for a photograph. Fax them interesting angles and quotes of participants.

Our Changing Society: Microcosm and Macrocosm

Loneliness

Many people say they watch TV because it "keeps them company." Perhaps in this hurried, mobile world we need to start taking time to develop more friendships with people of all ages and races. We can talk to them on the phone as we chop carrots for supper. If they live far away, we can write a letter. It doesn't require a lot of time. When we want a break from another activity, we can sit with our cup of coffee at the kitchen table instead of in front of TV. Writing a few lines a day eventually makes a wonderful letter. And who doesn't enjoy finding something personal in the mailbox?

> *Go without TV for a week. In this newly freed time, what can you do to let your spirit sing, test your resourcefulness, or infuse your life with exhilaration? If you do decide later to return to TV watching, note what jumps out at you.*

Even when there are several people in the same house, television increases the loneliness factor in our lives. With over two-thirds of American households having three or more TVs, often family members are in separate rooms watching their own favorite programs. Many parents have little idea what their children are watching.

Fear

Perhaps the defining characteristic of twentieth-century life is fear. Our perceptions of crime are generally much higher than the FBI statistics because of heightened media coverage. George Gerbner talks about "The Mean World Syndrome": the more action shows and news broadcasts you watch, the more you feel that the world is a dangerous place filled with evil people who are out to get you.[52] By and large we stay in our homes at night, remotes in hand, rather than venturing out to help someone in need.

Loss of Community

Being part of a community is not a matter of exerting effort to fulfill someone else's demands. Being involved with others satisfies what we all yearn for: the comfort of companionship and a sense that we belong somewhere and have something to contribute. We did not evolve as rugged individualists! (We did not evolve as rugged individualists but as supportive partners.) Our spirits soar in relationship with each other. While TV entrenches stereotypes, face-to-face interactions dissolve them. In real life we can delight in the delicious mysteries of each other. We need not merely tolerate diversity, but should celebrate it!

Pick a book you love or think is important and invite six to eight friends to read it within a month. Invite them to discuss it over coffee and dessert. (For fresh book ideas, see our bibliography.)

Loss of the Political Commons

Our passivity has deepened to the point that many Americans have given up the idea of having any say in the political process. Our political system has been so demeaned by politicians, pundits, and the influence of big corporate money that a shrinking number of people take any interest. In the 1996 presidential election, fewer than half of registered voters even bothered to cast a ballot. This is truly unfortunate, because decisions are made on all levels of government that affect our lives and our futures profoundly, whether we are conscious of them or not.

As Eric Alterman points out in *Sound and Fury: The Washington Punditocracy and the Collapse of American Politics,* much of the blame rests with the "squawking heads." "Pundits determine the content ... and set the parameters of our national debate.... Only they are accorded both the authority and the audience necessary to explain, in language a concerned citizen is likely to understand, why such and such proposal is either healthy or catastrophic for the Future of the Republic.... Though their editorials have nothing to do with objectivity ... reality is what the pundits declare it to be." Their impact is enormous, because television is the main information source for almost 90% of Americans.

Loss of Creativity

We live surrounded by an environment that pulsates *entertainment* and *escape.* But are our lives really so bad that we need to spend so many hours escaping? Do we expect everything to be pretty, easy, and comfortable, like a greeting card? In truth, life is very difficult at times, and pain is a part of growth. It's dangerous to use television to mask what we're really feeling. Spending hours and hours in front of TV keeps us from facing and feeling the problems that are our greatest opportunities for growth.

Whether plants or people, we all seek the light. In nature, the principle is *grow or die*—so, too, in our lives. If we are not growing, overcoming our procrastination, defusing our anger, or developing a talent, at some level we feel depressed and deadened. And we take it out on those closest to us. Growth and creativity are essential to our survival and our ability to nurture others.

Helena Norberg-Hodge gives a fascinating account of Ladakh, from which we can glean striking parallels to our life and times. Although Ladakh is tens of thousands of miles away on the windswept Tibetan Plateau, what happened to the Ladakhi people when television was introduced may help us see more clearly what TV is doing to us. Evening used to be the time for gathering and storytelling. Today, the Ladakhi people no longer tell stories, because their stories are not as polished and action-packed as those on TV. They no longer sing, because they can't sing as well as the man on television. They are embarrassed to dance, because they can't dance as well as the woman on TV. Tensions are increasing between old and young, men and women. Television is literally destroying their culture. What about ours?

Loss of Consciousness

People make jokes about their TV habit—"Yeah, I'm an addict"—but reality and jest are not so far apart. Just as drug addicts have to have their fix every day, those hooked on TV need theirs. Addicts lie to themselves about the seriousness of the problem and lie to get their fix. Kids often lie about having their homework done so they can watch the tube. A lethargic state is part of addiction—few people feel like jumping up and getting an important task done after watching TV for a couple of hours. Addicts have a hard time keeping promises, the contracts that hold our lives together. Our promise to our children is to nurture them. This takes time, for listening, comforting, disciplining. Our promise to ourselves is to figure out just what it is that's simmering below the surface—what is that special gift, talent, or passion that is uniquely ours to offer the world? When we spend a lot of hours watching television, there isn't much time left for promise keeping.

Remember a promise you made that you've been remiss in keeping. Take a step toward fulfilling that promise now.

Choosing Another Way

Hannah Arendt said "Evil is banal." Though we usually think of evil in large, horrific events like cult deaths, it more often slips in as the ordinary, commonplace, and trivial. *While we fritter away so many hours watching actors in their fantasy lives, we are missing the power and potential of our own lives.* Our heart is our real wisdom source. Often we need a little quiet space to sense what's inside us, to become aware of what our heart is urging us to do. Television steals both the inclination and the time for reflection.

Try an experiment. Don't turn on the radio, TV, or stereo when you come into the house. Let what you are really feeling rise to the surface. Perhaps you need to work on communication or healing in a relationship? Maybe you need to change your job or career. When we try to bury or run away from what we need to do, the problems don't go away. They resurface later with more force and more complications. The first truth the Buddha discovered was: "Life is difficult!" When we stop trying to avoid difficulties, when we look straight at them and feel the pain and confusion, we begin to see an initial step that can be taken. Weeks, months, or years later, we're grateful for the nugget of growth that came from confronting a problem.

Taking slow, quiet breaths while sitting outside can be helpful. Putting pen to paper can lead us deep inside. Starting with just one thought or question, the process of free-form writing often brings amazing clarity. Journals can become great friends.

When you have the urge to turn on the TV, instead try identifying feelings hovering just below the surface. Explore them in writing or by asking a close friend to listen to you without offering advice. Or engage in some kind of physical movement to shift your energy state.

Yes, we are tired at the end of a long day of work and commuting. But the paradox is that to receive energy, we must initially expend some of our own! I find that when I come home from work and just sink into the sofa, the evening goes down hill fast. But those evenings when I go for a walk, read with my daughter, fiddle with a project, or go out to a community meeting, I feel much better about life and myself and am able to cut others more slack. I have noticed that the most hopeful people are those deeply involved in creative projects or in actions with other people toward common goals.

We all need to blitz out in front of television at times. Occasionally, excellent programming allows the whole family to be entertained and uplifted. But television has assumed too large a role in our lives. For most of us, it takes the lion's share of our free time. *Choose life.*

It's the *time* we lose under TV's spell—the days, the weeks, the years—that's really disturbing. What difference will it make in a month or five years what happened on *Frasier,* or what the 49ers' final score was? Even with the higher-quality programming of PBS or The Discovery Channel, we are still getting our experience secondhand. *Choose life.*

In *The Evolving Self,* Mihaly Csikszentmihalyi writes about "flow," that experience we've all had—but too rarely—when the sense of time completely drops away. "Contrary to expectation, 'flow' usually happens not during relaxing moments of leisure and entertainment, but rather when we are actively involved in a difficult enterprise, in a task that stretches our physical and mental abilities.... Our whole being is focused in a harmonious rush of energy (which) lifts us up out of the anxieties and boredom that characterize so much of everyday life."[53] Any activity that's not passive can trigger flow: working on a challenging job, exercising vigorously, reading, or keenly focusing on trying to help another person. Television keeps us from experiencing it more often!

The stuff of our own lives is cooking, sharing meals, picking up after ourselves and others, helping with homework, getting ready for work, driving kids to soccer practice or music lessons, and going to meetings. It's teaching our children how to follow a recipe, re-attach a popped button, organize their closets—in short, how to live in the world. The fast pace and glamor of the entertainment media make these things seem mundane, so that it is much harder for us to get into the flow of making each day an exquisite masterpiece of pure intention, kindliness, and good effort.

Give yourself a big chunk of free time. Do an activity that you used to "get lost in." You may be surprised to discover that you still can.

When we extricate ourselves from hours of electronic media, we have the chance to connect anew with our families and friends. We can tune into the subtle nuances of moods and feelings. We have the time to ask the questions to draw forth the stories harbored within each of us. We can bring together an extended family of our own choosing—the wise retirees up the street, the families of our close friends, our favorite teachers or clergy—people we want our children to be able to lean on when we're frazzled and frayed!

Where do we go from here? Observing our own patterns and tendencies is one step. Helping our children structure their time more richly is another.

Making our voices heard in the network boardrooms is also part of the picture. And talking *everywhere* about the impacts of television—to raise a public dialogue—is essential.

At the end of Dr. Seuss's classic tale, *The Grinch Who Stole Christmas,* the Grinch waits in evil delight to see the crestfallen faces of the Whos in Whoville when they wake up Christmas morning to discover all the decorations, presents, and feast foods gone. But on that brilliant morning, he is instead stupefied to see the Whos gathering in the town square, holding hands and singing joyfully in the wonder and awe of the day itself. Wouldn't it be great if a Grinch unplugged a master TV switch, and we, released from the spell of television, spontaneously came together to laugh, sing, and revel in the beauty of the earth and of our own spirits?

Questions For Reflection

- Does the culture of disrespect so prevalent on TV, with its putdown language, sometimes infuse the behavior of your child or children you know?
- Does television stimulate a craving inside you for things you don't have? Are they essentials or luxury items?
- How do you feel after watching TV? Energized? Enervated? Elevated? Edified? Tinged with low-lying anxiety? Optimistic? Empty?
- What is your favorite TV show? What do you like about it? What does it bring out in you? Could you

use this insight as a springboard to a real-life activity?

- What have you been putting off doing or deciding while passing hours in front of TV? What promises to yourself or others are you not keeping?

Chapter 5

Media, Girls, and Body Image

Ellen Schwartz

how impossible images of physical perfection are making our girls sick

The tyranny of the ideal image makes almost all of us feel inferior ... We are taught to hate our bodies, and thus learn to hate ourselves. This obsession with thinness is not a trivial issue; it cuts to the very heart of women's energy, power, and self-esteem. This is a major health problem.
—Jean Kilbourne, in *Feminist Perspectives on Eating Disorders*

The anorexic is weak, voiceless, and can only with difficulty focus on a world beyond her plate.
—Naomi Wolf, *The Beauty Myth*

Many of our girls have "the look of sickness, the look of poverty, and the look of nervous exhaustion."
—Ann Hollander, *Seeing Through Clothes*

A plague is abroad in the land. It has not been caused by rats or mosquitoes, but by profit-driven industries relentlessly marketing an ideal body image impossible to attain. The victims of this epidemic? Our young women and girls.

Bombarded with Barbies

The onslaught starts with a seemingly innocuous toy. While Barbie is presented as fun and wholesome, her image sets the tone for what girls come to expect of themselves in real life. Her unnaturally thin body makes even little girls with normal bodies unhappy with their appearance.

Although my own daughter is supple and athletic, at age nine she declared she was fat because her tummy showed. Many first-grade teachers tell me about the lunches thrown in the trash because these girls are "on a diet." The Barbie message is amplified by advertisers: models who, twenty years ago, were 9% thinner than the average girl are now 23% thinner. Only 5% of the population naturally falls into this body type. For the majority of girls, this is a cruel impossibility.

The pressure of the body image is inherent in the doll; another part of the package is what children learn from playing with Barbie. Through play, children make sense of the world. The primary way girls play with Barbie is changing her outfits. Most of Barbie's clothes are not just stylish, but skimpy and suggestive. This emphasis on clothes leads to enormous pressures in pre-teen and teenage years to own sexy Calvin Klein jeans and trendy Nike shoes instead of generic brands at half the cost.

Barbie makeup kits lead girls to feel they're not really pretty without makeup, and set them up as eager consumers of the multibillion dollar cosmetics industry. And if that's not enough, the doll itself is made of polyvinyl chloride (PVC), and its manufacture and disposal creates deadly toxins.[54]

Girls as Disembodied Sex Objects

The early childhood playtime emphasis on the external (perfect body, alluring clothes) is the beginning of a continuum which for girls in our society flows seamlessly into the pre-teen years. In middle school, Barbie is supplanted by MTV. MTV not only showcases music videos but powerfully projects highly crafted images of relationships between the sexes. Ninety percent of the videos are directed by men, and the roles girls play are male fantasies. The girls are not presented as real people with hopes and dreams of their own. Camera angles focus on breasts and buttocks; often the faces of the young women are not even shown. Women are presented as having limitless appetites for sex with any available man.

I can hear some of you saying: "Oh, but that's just TV. Everybody knows that's not real life." Are you sure? In a survey of 1,700 seventh-, eighth-, and ninth-grade boys, the question was asked:

"When is it okay to force sex on a girl against her will?"[55] Twenty-four percent said it was okay if they had spent "a lot" of money on her ($10 to $15); 31% agreed it was all right if she had "done it" with other boys; 65% thought it was okay if they had been dating a long time (6 to 12 months). This survey seems to be borne out in real life: *32% of our young people have had sex by ninth grade!*[56]

MTV images are not the only ones fueling this behavior. Think about the familiar motif of television and movie stories (written mostly by men): Boy meets girl. Boy makes advances to girl. Girl is intrigued by boy, enjoys the kissing but rebuffs further sexual advances. Girl runs away. Pursuit ensues. Upon capture, boy forces a passionate embrace upon girl. Girl struggles momentarily, then sinks into sexual desire. Seeing this again and again, is it any wonder that boys get the idea that no means yes?

> *Watch MTV or a sitcom with a girl you love. During the program, or at the commercial breaks with the sound off, talk with her about how women and their bodies are treated (through clothes, words, and camera angles). Ask her how she feels about peer pressures and expectations regarding her body. Ask her what she wants in a relationship with a boy.*

Most young girls want someone to have fun with, someone who treats them sweetly and with respect. They want companionship. Older girls like hugs and kisses, but don't want to have to worry about getting pregnant. This reality is very different from the male fantasy we see reiterated on TV.

Fantasy Images Are Making Our Girls Sick

A homogeneous culture is sweeping our country, created by the mass media. Prettily packaged but with harsh undertones, it endangers our girls both in urban areas and in the heartland of America. Mary Pipher, author of *Reviving Ophelia,* is a psychologist in Lincoln, Nebraska. For twenty years, her focus has been on girls and their families. At first, girls came to her because they felt fat or ugly, or hated their parents. The four principal reasons why girls now seek her help are anorexia, bulimia, gang rape, and self-mutilation. Why are there such enormous pressures on our girls? How is it they find themselves so besieged that they feel nothing except when they're scarring their bodies?

What has happened to childhood? What has happened to adolescence as a time when girls are testing their wings, delighting in the vistas of the world opening up to them? Courage and exhilarating adventure should be the hallmark of their days as they explore budding talents and hone their skills.

Compliment girls you know on their achievements to emphasize that these are more important than how they look. Encourage non-stereotypical goals, roles, and dreams.

We have stood by as the commercial media, with technological glitter and glitz, reduced the acceptable female body size by one-quarter. The misery deliberately engendered by the pervasive marketing of this impossible body ideal is incalculable. The diet business has tripled in the last ten years, from a $10 billion to a $33 billion industry. Yet of those who drop twenty-five pounds or more on a weight loss diet, 90% regain the weight within two years, 98% within five years.

The natural physiological increase of body fat during adolescence is now viewed by many girls as nasty. Their attempts to force their bodies into the unrealistic ideal can lead to severe dieting and semi-starvation, often with diet pills. The negative consequences are many. Psychological effects include irritability, poor concentration, chronic anxiety, depression, apathy, mood swings, insecurity, fatigue, and feelings of social isolation. Physical risks include heart and kidney damage, hypertension, stroke, rotting of teeth (from vomiting), disruption of the menstrual cycle, loss of skin tone, lack of mental clarity, and decreased energy.

Bringing Balance with Better Role Modeling

The mass media will not tell the whole truth about this story. It would cut into sponsor profits and corporations would pull their advertising dollars. It is up to us to make this a prime topic of conversation with all of our girls—daughters, granddaughters, nieces, neighbors, friends' children. We must help them gain perspective on the tyranny that puts such crushing pressure on them.

To help our girls find their natural weight, we must focus on balance and wellness, not a number on a scale. We must emphasize eating nutritious food—centered around fruits, vegetables, and grains—in a relaxed and satisfying manner. We must support them in keeping movement in their lives, such as gymnastics, ballet, hiking, swimming, or gardening.

What gesture could you make to a girl within your sphere to help her bring more balance, health, and wholeness into her life?

Naomi Wolf (author of *The Beauty Myth),* points out that the paring down of female body size intensified in the 1960s. When women's houses no longer confined them, their bodies became the new prisons.[57] Many of our young girls' attitudes are absorbed from their mothers' experiences. How much are we still buying into this hype? In addition to working full-time, getting the laundry done, helping with homework, schlepping the kids around, and holding the family together emotionally, must we women also struggle for a great figure? *Do we put even one-tenth the time into bringing about societal change as we do into diet strategies?* Perhaps if our daughters saw us working to stop the rollback of wages and healthcare benefits, or attempting to halt the destruction of the ozone, they would realize that having a perfect body is a trivial pursuit.

Initiate an action with a young girl on an issue of mutual interest that goes beyond your own lives, such as attending a rally against land mines, speaking up at a school board meeting about the use of pesticides on school grounds, starting an organic garden, writing a letter together to an elected official, or...?

Questions For Reflection

- Do you have memories of being made to feel "less than" as you grew up seeing advertising and movie images? Are you comfortable with your body now?
- How do you think the messages sent by the fashion and diet industry affect our girls? Think about girls you know or have known.

- What was something wonderful about your childhood and adolescence that girls rarely get to experience today?

Chapter 6

The Best Government Big Money Can Buy

Ellen Schwartz

Can a corporate-sponsored democracy serve the people?

We must crush in its birth the aristocracy of our moneyed corporations, which dare already to challenge our government, and bid defiance to the laws of our country.
—Thomas Jefferson

If not now, when? If not us, who?
—Robert Kennedy

Before I ran for congress in 1994, I thought I was politically savvy. But what I learned about our political system from the inside made me grieve for this country. The glitzy campaign appearances and glib pundits lull us into thinking that we have a functioning democracy, but we do not. It is not only Third World countries that have sham democracies. I would love to share with you the amazing journey I was privileged to take as I came to this view of how our democracy really works.

As the daughter of a carpenter and a department store salesperson, I never wrote essays in high school about how I planned to be President someday. Though I was interested in both domestic and foreign affairs, the idea of holding political office was so far out of my ken as to seem unreal.

Saying 'Yes' to the River

Periodically, I would volunteer in campaigns and initiatives. In 1992, my district's congressional race caught my attention. A farright Republican from the state assembly was running against a rather lackluster Democrat. As I was resignedly stuffing envelopes the week before the election, I commented to the other volunteers around the table, "If we're ever going to get that right-winger out of office, we'll have to get a dynamic woman to oppose him." From who knows where, the words flashed through my mind, "That's you, honey!" Of course, I kept my mouth shut!

My son was then fifteen and my daughter six. My husband and I were running an equipment rental company employing twenty-seven people, which we had started fourteen years before. My "spare" time (after grocery shopping, cooking meals, doing laundry, keeping the yard up, and driving kids to lessons and sports practice) was focused on volunteering for environmental issues. The growing hole in the ozone and massive destruction of rain forests struck an urgent chord with me. Well into 1993, I put all thoughts of political office far out of my mind. Mid-year, however, I would find myself waking up during the night plagued by the thought "You should be running for Congress!" A flyer from the National Women's Political Caucus appeared in my mailbox, promoting their upcoming workshop for women considering running for office. I gave in. I signed up for the weekend.

Down to my bones, I realized that the incumbent congressman, Bill Baker, opposed nearly everything I had ever worked for or believed in. He considered the environment to be raw material for corporate profit; I passionately believed pollution and over-reliance on fossil fuels were seriously compromising the world we were leaving to our children. He wanted to abolish the Department of Education; I firmly believed that the public education system was an important element in the strength of our nation, and the current crisis meant that it needed a vast infusion of new ideas and resources.

Baker was anti-choice on abortion; I supported a woman's right to have the ultimate control over her body and her life. Bill Baker's policies advantaged multinational corporations and his campaigns were fueled by contributions from their Political Action Committees (PACs); I fiercely believed that small and medium-sized businesses should be helped most by government. It is they who provide the greatest number of jobs and, quite often, the more satisfying and less stressful ones.

I sensed a rare opportunity. The realization seemed to be spreading that women offered a different perspective on public office, and more and more women were daring to jump into the process. Not only did I have a background in small business and ecology, but I had a Masters Degree in Speech and Communication Studies. I was also experienced and knowledgeable in the area of Soviet-American friendship and business exchanges, having been a founding board member of The Center for US-USSR Initiatives in the mid-1980s.

I felt that many people would find my background and perspective a refreshing contrast to those of the incumbent. I also felt that our nation was at a critical turning point. The "peace dividend" had already been siphoned off; more and more families were under extreme duress as their chief breadwinner lost a living-wage job due to corporate flight to cheap Third World labor markets; and healthcare was out of reach for a growing number of people.

Our youth are very perceptive. Some saw what was happening, and redoubled their efforts to get into the best schools, to secure the best jobs. More watched the growing tide, and their hope trickled away. What was it replaced by? Alienation. Rage. Despair.

I really thought I could make a difference. I really thought I had a good shot at winning the election. I could not let our children down—*all* our children. In any event, I knew I could not face my own children's probable questions in ten years ("Why didn't you seize the opportunity to make things better?") if I did not give my all to win this office.

I'd spent a lot of years of my life being a worrier and spending my money very cautiously. The moment I said yes to seeking congressional office, however, I felt that I was connected to a wide, deep-flowing, slow-moving river in the earth's center. I had never before experienced such a profound sense of calm and certainty. I had no idea what was in store, yet I was worried about neither small details nor large events.

A Leap of Faith

The consultant I hired said I needed to "seed the environment" with a broad-based mailer to the "movers and shakers" and "opinion-makers" in my area. Initial cost: $10,000. I who always bought the generic brand to save twenty cents took a big breath, felt the deep, slow-moving river, and signed the check. Then she said I needed to "scare off any other potential contenders" by stating in the letter that I was prepared to spend $50,000 of my own money to win this race. I was shocked! My husband, Michael, and I had worked such long hours to drive our business forward, and though it had done better than we had ever expected in our wildest dreams, $50,000 was a mind-boggling amount for us. I thought "Oh, why not, we won't really have to put up that kind of money." Little did I know!

I did have one challenger in the primary, but I beat him 62 to 38%. I was exhilarated. Now the tough part of the race was to begin, against the incumbent who had held state and federal office for a total of fourteen years. I knew it would be an uphill battle, but he was rigidly conservative and this was a moderate district with a highly educated constituency.

When running for national office, either as a Democrat or Republican, the party's Central Campaign Committee (CCC) is essential for survival. No person can possibly raise enough funds to run a campaign in these days of media saturation without the data input and monetary help of the central party. Before I won the primary with such a large margin, the DCCC refused to believe that I could be a realistic contender against Bill Baker. They were still skeptical. They were, of course, willing to supply briefing sheets on issues and hot-button events, but would not give me any money unless I matched them dollar for dollar!

I remembered what my salt-of-the-earth mother had said to me months earlier when I told her I intended to run for office: "Our people don't do that kind of thing! Politics is a rich man's game!"

The year before, about the time I initially decided to seek the congressional seat, my husband and I had sold our business to our partner. Now he, too, wanted to sell, but the buyer would not purchase the business without the land. Michael and I had no other savings or retirement account, and wanted to keep the land for our later years. Under pressure, however, we acquiesced. Just at the juncture when the DCCC would not invest in my campaign without personal matching funds, we had liquid money. Putting up the $50,000 was a decision we now had to make!

Michael had always believed in me and was my strongest supporter in running for office. Now he said, "Go for it!" The squirrel in me, who wants to bury all the nuts for the long cold winter, said "Are you insane? You've never seen this kind of money before, nor is it likely you'll ever have this amount again! Save it!" The other side of me, drawn down to the deep river, said "This is part of the process. This is the right path for you to be on now. Do it."

I did it.

Money and Mine Fields

The personal money was one hurdle; the bigger challenge was raising money from others. I had, until then, been afraid of fundraising. Isn't it interesting how life always puts directly in front of us what we're trying desperately to avoid! To my surprise, I found that asking for money from individuals wasn't really so bad, once I got started. By the end of the campaign, an enormous number of "regular people" (more than 2,100) had contributed. But it is impossible to run a viable campaign with $25 and $50 checks. Since I knew very few people who could afford to give the legal individual limit of $2,000, I had to go after PAC monies.

I was happy to get major contributions from educational, environmental, labor, and women's PACs because I was sympathetic with their needs and issues. Unfortunately, these groups have far fewer funds than chemical and oil companies, the telecommunications industry, and transnational corporations. In addition, most PAC money (about 90%) goes to incumbents. My opponent had the great advantage.

But it was much more complicated than that. Even groups with which I had a lot in common, such as Jewish PACs, did not give me funding support because we were not in agreement on all the issues. As a Catholic woman married to a Jewish man, I understood their mindset, but I do not believe that the Israeli Jews have dominant rights to the Holy Land. The

Palestinians, too, have lived there for millennia. I see the struggle in Palestine as the archetype for many claims and conflicts around the world. None of us is totally right, nor do we ethically have full claim to what we desire. Peace will not last unless it is crafted in partnership.

Getting $10,000 contributions from PACs is one struggle in a campaign; getting in-depth coverage from the media is another. The print and television media are the lifeblood of a political race; only if you are a multimillionaire can you afford to buy enough time to elevate your race in the public mind. One key issue in my district was the Lawrence Livermore National Laboratory. I was a strong proponent for keeping the best and brightest minds assembled there working, but not on nuclear weapons and nuclear simulation technologies.

I loved Energy Secretary Hazel O'Leary's idea of making Lawrence Livermore the prototype Green Lab in the nation and the world. The environmentally friendly and economically feasible projects that could be developed include rapid transit; energy-efficient heating, lighting, and insulation products; and technologies to detoxify industrial waste and transform it into something useful. Of course, shifting the focus to more ecologically benign goals would involve some temporary dislocation, and a few would lose their high-paying jobs. But retraining was an important part of my proposal. I would have loved to have even one article or a five-minute TV news segment to explain these ideas, but I was never given the chance. The owners of the major print and commercial television media do not believe that complicated issues hold the public's interest and attention.

Instead, I was pegged as "Anti-Lab" because I did not endorse keeping every last job, and because I did not support the latest project for which the Lab was seeking congressional approval and funding: the National Ignition Facility (NIF). NIF's construction costs were then estimated to be $1.2 billion, with life-cycle operating costs of $4.6 billion (costs of decontamination and decommissioning would be extra). While the Lab's press releases stressed that NIF advances laser fusion technology, many of the forty-four nations of the world being pressed to sign the Comprehensive Test Ban Treaty are concerned that it will allow the United States to keep our current stockpile in peak readiness, while designing and testing key fusion aspects of the next generation of nuclear weapons.

Do you want to lend your efforts to stopping NIF and keeping the integrity of the Comprehensive Test Ban Treaty? Call Tri-Valley Communities Against a Radioactive Environment (925-443-7148).

It's All in How You Frame It

The political process is ugly, not only for keeping the sound-bite focus squarely on short-term gains, but also for bringing the temper and tone of public debate down to the lowest common denominator. Candidates are continually advised by their professional consultants to be as vague as possible so as not to alienate anybody. I was admonished to water down my comments so I wouldn't be labeled "anti-defense" or "liberal."

Eleanor Roosevelt's observation is perhaps even more applicable now than when originally stated several decades ago: "It's amazing how radical pure common sense is made out to be!" A major reason I decided to run for Congress was to help set U.S. funding priorities on a more sane path. Although the Cold War is over, the United States has barely reduced its annual defense spending to $270 billion. This is more than the combined military budgets of the next ten most heavily armed nations—and all of them either are or want to be our friends! Russia spends less than $50 billion, and her poorly paid soldiers are often seen siphoning gas from tanks to trade for food.

While defenders of the military budget claim that we spend only a quarter of our national budget on defense, this is a carefully framed lie. When the military budget in the days of the Vietnam War began to take a lion's share of national revenues,

President Johnson's administration came up with a way to hide the massive expenditures. Previously, Social Security funds were not counted in the general budget. Social Security monies, accounting for about half of the funds in the national treasury, are collected separately from income tax and are disbursed only for that purpose. Social Security money is not part of the discretionary budget.

When Social Security is falsely included in the budget, it looks like defense only gets about a quarter of national funds. If instead we look at the money Congress actually has the power to allocate, defense gets over half the budget. This includes expenditures on weapons design and procurement, personnel, domestic and foreign bases, and interest on past wars. As you drop your income tax payment into the mail, think of more than half of each tax dollar going to the military!

This excessive spending on the military is contrary to the desire of the vast majority of Americans. Polls consistently show that consensus reaches across gender, regional, and political lines: 77% feel that the budget for nuclear weapons is too high and 87% want the United States to negotiate an agreement to eliminate nuclear weapons.

Labeling is a quick and easy way to destroy a candidate. Though reducing the military budget is a matter of common sense, being labeled "soft on defense" is still very deadly in a moderate-to-conservative district.

I was also labeled "soft on crime" because I did not endorse prison construction as a top priority. Instead, I talked about the basic reasons why people are driven to crime: poverty, poor schools, and lack of living-wage jobs. Inner-city unemployment statistics are more than three times those of primarily white suburbs, and this is before taking into account the cruel deception of the unemployment figures. As Michael Parenti notes in *Democracy for the Few,* if you are not currently collecting an unemployment check, you are not even counted in the official numbers! Those whose benefits have run out are simply not part of the statistics. The actual unemployment rate is at least double the official number, and when you include the number of people who are only able to find part-time work, the real unemployment rate is probably closer to 14 or 15%. These are not concepts that can be readily explained in a nine-second sound bite.

Pick up a publication that adds alternative views to the spectrum of opinion (YES! A Journal of Positive Futures, Atlantic Monthly, Harper's, Mother Jones, The Nation, The Progressive, Utne Reader). *Or, tune into the Pacifica Network on FM radio—stations that accept no corporate monies. (See* Resources *at end of the book.) What differences do you notice?*

Malls and Media Undermine Democracy

Today's mainstream media does not foster a healthy democracy. Candidates are not given in-depth coverage, nor are issues presented from multiple points of view by the select talking heads. The debate ranges from centrist to conservative right. We hear the same news bites ad nauseam.

Even a candidate's ability to meet people face to face has been seriously diminished. The new commons are the shopping malls, farmers' markets, and wine and art festivals. These are privately owned, and most have restrictions against political candidates talking with their customers or giving out materials. Yet where else do people gather? When I did walk these events, I was made to feel like a criminal. I had to give literature to people furtively, so as not to be seen by security and thrown off the premises. The offense? Distracting customers from spending money.

It is true that the League of Women Voters sponsors candidate forums to foster exchange of ideas. The trouble is, almost no one shows up for these debates. It is perfectly understandable—people are tired when they come home from work. They have meals to cook, children to feed, bathe, or help with homework, and preparations to make for the next day's work. The debates are broadcast, but generally only on local cable channels whose program schedules do not even appear in the printed TV listings.

The logical way for people to get a sense of the issues would be for the major networks to devote one hour of prime time each evening for the six weeks before the election to covering a different initiative or race at no charge to the candidates' campaigns. After all, the airwaves are our public property, and they are leased to the networks at no charge *in the public interest.* All the rest of the time they could continue to rake in the multimillions from the advertising they sell.

> *If you want to be an active online citizen, join Public Citizen's cybercitizen network at* www.citizen.org.

Despite raising almost $450,000, I was not able to get my message out broadly enough to defeat my opponent. Ellen Tauscher, who defeated Bill Baker two years later running on many of the same issues, mounted a $2.4 million campaign, putting in $1.67 million of her own personal fortune. This gave her the ability to blanket the market with two straight months of TV spots (which I had not been able to afford at all). Having built her wealth as a stockbroker, her Rolodex included people who could make far bigger contributions than my friends and acquaintances.

Democracy for Dollars

We no longer have a viable democracy. Only if you are wealthy can you amass the enormous funds necessary to create the high visibility campaign that gives you a chance of winning. It does not matter if you are bright, energetic, hard-working, and knowledgeable on national and international issues. Courage, compassion, and commitment are not enough to win an election. Money is the most reliable indicator of who gets into office. *I was outspent by Bill Bakertwo to one and he won; in the following election, he was outspent two toone and lost. This holds true in over 95% of elections.* No wonder more than 60% of people who are eligible to vote choose not to, even in presidential years.

Connect with local members of Common Cause and help forward their initiatives on limiting large corporate and individual campaign contributions. Call 202-833-1200 or look them up at www.com moncause.org.

The demise of our democracy has happened because big money is calling the shots. The most effective step we could take is public financing of political campaigns. Unfortunately, this is still an unpalatable idea for most Americans because of negative media campaigns financed by those who stand to lose their power. Campaign contributions are given for the purpose of creating legislation favorable to special interests. Yet most people still do not connect the dots about the direct consequences to the public good: lost tax revenues, unethical tax breaks, environmental deterioration, and policies that promote job loss in this country.

Meanwhile, we can begin to level the playing field by:

- Mandating that major networks give one hour of prime time each evening for in-depth coverage of races and initiatives starting six weeks before the election.
- Limiting the amount of money that can be spent by candidates: $200,000 would be adequate for a congressional race, $2,000,000 for the Senate.
- Limiting the amount PACs can give individuals to $1,000.
- Limiting the amount of "soft money" that can be given to central party committees to $1,000.
- Lowering the amount of individual contributions from $2,000 to $100.

Late on election night, when it was clear that Bill Baker had been re-elected, I was surprised but I didn't feel defeated or a failure. Through the tears of exhaustion and release, I had a strange sense that all was well.

Do I have residual regrets about the experience? Not at all! I couldn't have gained such a poignantly clear picture of the inner workings of our current political process if I'd gotten a Ph.D.

I came to deeply understand that we *cannot afford the luxury of not being politically active.* Yes, it is harder to decipher a ballot these days—to really know who you're voting for and the consequences of all the initiatives. Though it takes more time, attending forums where candidates are present and initiatives are debated is still the best way to become informed. And, women and men must both take up the challenge of running for office. When a problem deeply concerns us, we must be willing to step to the fore and become immersed in figuring out the solution.

For women, running for office is a tougher choice. If they are mothers, they must still meet the many needs of their children. (I was fortunate that my husband was willing not only to help me on the campaign but also to take over most of the household and childcare responsibilities.) In addition, women generally know fewer people with ample checkbooks who can help fund initial efforts. Running for local office is a good place to start—less money is needed to wage a campaign and you garner the name recognition helpful in state and national bids. Persistent grassroots effort can still create surprising results. Undeterred by a far better funded and well-known candidate, Audie Bock of the Green Party won a California Assembly seat in 1998 by door-to-door walking and talking.

The office itself isn't the only thing to win. As the weeks went by and I realized that my next task was in the nonprofit world, I saw that the people I met in the campaign would be my key to success in helping youth. When I needed high-profile people to be on the Advisory Board of newly forming *Invest in Kids,* one telephone call was all it took. When I needed volunteers to be mentors to the low-income, at-risk youth we were focusing on, many of the people who had worked on the campaign plugged right in. When we needed media publicity to reach out for scholarship donors, press releases sent to newspapers, radio, and TV were responded to with wonderful stories because I had name recognition.

Call the National Women's Political Caucus (NWPC) to get the phone number of the chapter nearest you and get on their mailing list for upcoming workshops about running for office: 202-785-1100. Address: 1630 Connecticut Ave. NW, Washington, DC 20009, www.nwpc.org.

Whether running for office, helping on a campaign or initiative, or doing the down-in-the-trenches work of making calls, writing letters, and spreading the news about pending legislation, we must all be in this for the long haul. Decisions that affect our personal lives, our work lives, and the viability of the planet itself are being made on all levels of governance. No matter how small or large our part, it is imperative that we do *something.* Though it is easy to feel separate, we *are* a part of a seamless whole. What we do or let slide does matter, does make an impact, whether we see it immediately or not. We absorb as if by osmosis the desire for immediate gratification, yet the rhythm of life is long and measured, with many harmonies blending. Is this serious? As life and death. Is there room for lots of fun along the way? You bet!

Call the candidate or political party headquarters of your choice. Don't forget alternatives such as the Green Party and the New Party, which are winning local and state races across the country, or the Natural Law or Labor Parties. (See Resources *at the back of the book for phone numbers to connect with local affiliates.)*

Questions For Reflection

- What aspects of our political system seem least balanced and wholesome to you?
- How would you define the responsibilities of citizens in a democracy?
- What is one step you could take to add momentum to positive change?
- What is your highest vision for our country (vividly imagine political rejuvenation, economic account-

ability, wages and work hours, defense expenditures, spending on healthcare and education, the prison system, care of the environment)?

Chapter 7

What Do World Trade Agreements Have to Do with Me?

Ellen Schwartz

globalization means equalizing down to the lowest common denominator

Under GATT, the race to the bottom is not only in standard of living, environmental and health safeguards, but in democracy itself. Enactment of the free trade deal virtually guarantees that democratic efforts to make corporations pay their fair share of taxes, provide their employees a decent standard of living, or limit their pollution of the air, water, and land will be met with the refrain, "You can't burden us like that. If you do, we won't be able to compete. We'll have to close down and move to a country that offers us a more hospitable climate."
—Ralph Nader and Lori Wallach

Never doubt that a small group of committed people can change the world. Indeed it is the only thing that ever has.

—Margaret Mead

Life as we know it has dramatically changed, but most of us still drive the kids to school and hurry to work without a clue.

GATT, the General Agreement on Tariffs and Trade, passed the U.S. Congress in the fall of 1994 after a mere twenty hours of debate. With that vote, our ability to protect the environment, our health, and our local economies was signed away. It is now a matter for three trade bureaucrats in Geneva, Switzerland to decide if our fuel efficiency mandates, pesticide residue laws, and statutes for recycled content in newsprint are "nontariff trade barriers and therefore illegal under GATT." Can't believe it? As a matter of fact, the dolphin-safe tuna law that so many Americans worked for has already been declared GATT-illegal. We are now importing tuna caught using mile-long nets placed around schools of dolphins. Does this kill your appetite for a tuna sandwich?

Buckle your seat belts for a short but terrifying tour of GATT and its administrative arm, the World Trade Organization (WTO). We'll also expose you to the next level of treaties coalescing at the world trade talks in Seattle in early December 1999. On the docket of this "Millenium Round" is the Multilateral Agreement on Investment (MAI), intellectual property rights, health and educational services, and a global free logging agreement.

Who Is GATT Good For?

Most of us have heard the phrase "What's good for General Motors is good for America." Once it seemed to fit right in with the national psyche. There actually was some truth in it a couple of decades ago. Those were the days when many people were getting a good salary (the country being 33% unionized had a lot to do with overall worker gains). That was the time when parents could afford to help their children through college, and when they knew their kids were going to have a better life than they did. And though we saw Ohio's Cuyahoga River burn on national television, we were confident we were going to get a handle on pollution with the new Clean Water and Clean Air Acts.

GATT is very good for General Motors, Exxon, Dupont, Wal-Mart, IBM, Philip Morris, and the other Fortune 500 CEOs and stockholders. But it is neither good for the people of the United States nor for those living in Third World countries.

Herman Daly, the economist, notes that the more accurate name for this "free trade" agreement is "deregulated international commerce." Under GATT, countries may not give support to their own small, local companies for contracts; tariffs and import quotas must be dropped on all goods and agricultural produce; and "nontariff barriers" such as regulations protecting health, workplace safety, and the environment must be eliminated.[58]

Future Shackles

In *The Case Against the Global Economy,* Ralph Nader and Lori Wallach point out that, unlike previous trade agreements, GATT requires all participating countries to be bound by all agreements. The new rules supercede all existing and *future* federal, state, and local laws. Decisions about compliance are now made by the World Trade Organization, the enforcement arm of GATT. Its rulings are automatically binding unless *all member countries* vote to stop the decision *within 90 days.*[59] Can you imagine *anything* that a hundred countries could agree on—let alone in a 90-day period?

In the past, government purchases have furthered policies such as recycling or alternative energy development. Governments can no longer do this—they must now treat foreign and domestic companies equally in securing goods and services. In addition, other countries may now challenge a nation's *internal* laws. United States laws targeted for repeal by Europe, Japan, and Canada include the Delaney clause (prohibiting carcinogenic food additives), the asbestos ban, driftnet fishing and whaling restrictions, the Corporate Average Fuel Economy (CAFE) standard and gas-guzzler taxes, state recycling laws, and the Clean Air Act. In fact, *every environmental or public health law brought before the WTO to date has been ruled illegal; and 80% of our environmental legislation could be challenged!*[60]

Secret Decisions Against Hard-Fought Gains

Who in the WTO is actually empowered to declare our democratically created laws illegal? A secret committee of three nonelected trade officials is making the binding decisions. Only representatives of national governments may be a part of the hearings; citizens, nongovernmental organizations, the press, and even state and local officials are locked out. If our laws are declared GATT-illegal and we insist on keeping them in effect, we will have to pay fines in perpetuity to the nations protesting our safeguards. *Instead of raising world-wide environmental, health, and labor standards, GATT brings them down to the lowest common denominator.*

One recent example is the U.S. law to protect endangered sea turtles. The American law required shrimp trawlers to use Turtle Excluder Devices (TEDs). WTO ruled the law illegal under GATT and decreed that all shrimp must be allowed into American markets. The basis of this ruling? WTO policy: "Regulations designed to influence the regulatory regimes of other countries are unjustified discrimination against free trade." It is enormously disturbing that GATT/WTO dispute panels ignore multilateral environmental movements and agreements. As Public Citizen, the nonprofit consumer advocacy organization founded by Ralph Nader, notes: "One of the key components in setting the world on a sustainable and equitable development path involves changing the conditions and processes under which goods are produced and commodities grown, harvested, and extracted."

Of course, GATT rulings cut both ways. Out of concern for potential health hazards, the European Union (EU) did not allow beef cattle, domestic or foreign, to be treated with growth hormones. The American beef industry pressured the U.S. government to challenge the EU ban: the growth hormone ban was declared to be in violation of GATT, and Europe was forced to import American beef. Three unelected trade officials thus put themselves in the position of final arbiters of safety standards, overturning local and national laws created to protect citizens' health and well-being.

Who's Standing Up for Our Quality of Life?

How could a trade agreement so detrimental to our quality of life have been approved by Congress and signed into law by the President? Given the narrow and superficial quality of media coverage during campaigns, it is impossible for a candidate—even an incumbent president—to get his message across without massive monies from large corporations. That GATT was pushed through by Bill

Clinton is testament to the enormous power of the transnational corporations.

The same transnational corporations that fill political coffers, own the media, and dominate our economic lives. The deliberate decisions of corporations which were formerly rooted in the U.S., then grew to multinational size, and have now expanded to transnational status beyond the power of any government are at the heart of our economic insecurity.

Because these transnationals own the major television and print media, they determine what is news. We see cute dancing suns and rain clouds during the TV weather forecast but we are not told that nine of the hottest years on record were within the last twelve years—a strong indicator of global warming, with its potentially drastic impact on climate.

Overt incidents are reported (like accidental toxic gas releases from refineries) but there are no probing stories about research that links organochlorines to breast cancer. Organochlorines, such as dioxin, PCBs, and vinyl chloride are byproducts from the production of chlorine that are used in pesticides, plastics, and solvents. In the United States, breast cancer is now increasing by 4% annually, and women with high concentrations of these chemicals have breast cancer risks 4 to 10 times higher than women with lower levels.[61] It is noteworthy that there is a Clorox factory located in Oakland and that the San Francisco

Bay Area has one of the highest rates of breast cancer in the nation.

> *Organochlorines persist for decades, even centuries, and accumulate in the tissues of living things.*
>
> *• Contact the news department of your favorite television station, and ask for stories about the impact of organochlorines on public health.*
>
> *• Work with Breast Cancer Action (415-243-9301).*

The Fortune 500 corporations fund political campaigns, not out of civic duty, but because they expect something in return. *More favorable conditions for big business than those provided by GATT could hardly be imagined.* Do we really think that American consumers benefit when corporations avoid paying living-wage jobs in the United States by setting up operations overseas where workers are paid only 29 cents per hour? Nike's shoe production in Asia is an incredibly profitable venture for shareholders, but not for parents harangued to buy $129 basketball shoes.

In fact, the American trade deficit has grown in the four years since the WTO was approved. The U.S. trade deficit increased from $104.4 billion in 1994 to $113.7 billion in 1997, with the first two quarters of 1998 being 44% higher than 1997. Trade deficits translate to more jobs being destroyed by imports than are created by exports, and the Economic Policy

Institute found that the better-than-average jobs were the ones eliminated.

The Fallacy of the Global Trade Myth

While world trade is being promoted as the answer to our problems, the facts tell a different story. From World War II to the mid-1990s, the world economy grew fivefold and international trade increased twelve times. Yet more people live in desperate poverty now, both in the Third World and in America; violence is increasing within families, on the streets, and between nations; and ecosystems are being depleted beyond regenerative capacities.

The power of transnational corporations has been vastly enlarged by GATT. Corporations now have absolute freedom to set up operations where wages are cheapest and environmental regulations are the least restrictive, either by pitting localities against each other domestically, or by establishing operations in Third World countries. There is a big difference between *price* and *cost.* The *price* is what a purchaser pays; the *cost* is what is borne by communities whose local governments must pay emergency medical bills (for those who can't afford health insurance) and clean-up expenses for toxic wastes left behind by vacating businesses. GATT gives corporations greter license than ever before to externalize the costs of making their products. Is it worth it to keep stockholder dividends high and CEO compensation packages in the multimillions? Our corporate-owned media doesn't talk about who the *real* winners and losers are.

Justifiable Outrage

Perhaps we can take hope from how GATT was passed. GATT was not translated into the native languages of most countries; the leaders and legislators approved the 700-page document with only summaries provided by the agreements' promoters. Nonetheless, the people in many countries took a much more active stand against GATT than Americans did. Nader and Wallach point out that outraged citizens had to be dragged out of the Parliament in Belgium before the vote; the opposition in Spain was so strong that the vote was hastily scheduled with a skeletal parliament on Christmas Eve; and there were anti-GATT street riots in the Philippines and in India.

In India, the people so fiercely opposed the intellectual property rules (which would make them pay royalties to transnational seed companies on plants they and their ancestors developed) that they forced their Parliament to eliminate approval of the portion of the treaty creating the enforcement arm, the World Trade Organization. Because the world treaty was supposed to be approved with no deletions or amendments, the Prime Minister reinstated the WTO section by executive decree. Six months later the Parliament, standing up for the people, vetoed his action.

If farmers in India could fight for their rights so effectively, why not educated Americans? This nation was created by people who rejected the rigid, illegitimate power of aristocracy. The struggle about GATT may be in abeyance, but *we imminently face the challenge of a new structure that will dominate our lives even more completely: the proposals first put forward as MAI.*

MAI: The Cancer Grows

MAI was initially crafted by an elite group of twenty-nine wealthy nations in regular consultation with their corporate lobbies. America's elected leaders, as well as labor, environmental, and community organizations, were kept in the dark until the treaty was almost fully formed. Public opposition was so swift and fierce that MAI was withdrawn as a treaty proposal in December 1998. The victory was short-lived, however. Now several of its tenets are being inserted into new global trade treaties, and the MAI itself is on the agenda for incorporation into the WTO at the "Millenium Round" of World trade talks in Seattle.

Whereas GATT applies to trade in goods and services, MAI pertains to investments. For every dollar that circulates in the productive economy, $20 to $50 circulates in the world of pure finance. The wealthy are not making money by creating more goods and services, but by moving their funds around the world with lightning speed.

MAI was designed to release corporations and investors from the restrictions that national governments can place on them. If its proposals are enacted, the MAI would severely limit our ability to protect labor rights and the environment and to promote social and economic justice. *The MAI places corporate profits, present and future, above all other concerns.*

Foreign corporations could sue the federal government over laws that might limit corporate profits, such as environmental, labor, health, human rights, or local community development standards. This would apply to both present and future profits. The governments would have to overturn their laws or pay multimillion dollar damages.

Does this seem unbelievable? It's already happening! MAI draws heavily on NAFTA, the North American Free Trade Agreement, passed in 1993. Under a NAFTA provision similar to MAI, the U.S.-based Ethyl Corporation sued the Canadian government and won $251 million in damages related to a health and safety law. The law banned a gasoline additive, MMT, produced exclusively by Ethyl Corporation. Despite the fact that MMT has been banned in many U.S. cities because it harms cars' pollution-control systems and has toxic properties, Ethyl claimed it should be compensated by the Canadian government because its law would hurt potential future profits and harm Ethyl's reputation![62] *If MAI tenetsare injected into the international trade treaties currently being negotiated, no local, state, or congressional legislator could pass a law without worrying about the potential for foreign corporations suing them for cash compensation—taken from taxpayer dollars.*

More Bad News

What else would happen under MAI-type proposals? The *Ad Hoc Working Group on the MAI* points out that they would prohibit governments from requiring foreign investors to maintain an investment in a community for a set amount of time, use recycled or domestic content in manufacturing, or hire local workers, even if the same requirements are applied to domestic companies. They would ban governments from regulating business operations to control environmental damage, or offering priority to companies using the best environmental practices.

These proposals would prevent governments from nurturing native industries, small businesses, or women and minority-owned companies by being forced to offer the same advantages to foreign corporations. They would prohibit countries, states, or localities from using human rights, labor, or environmental standards as investment criteria. Had these proposals been in effect in the late 1980s and early 1990s, the investment strategies aimed at abolishing apartheid in South Africa would have been forbidden.

MAI-type tenets make it much easier for corporations to move capital where it is most profitable and least accountable, accelerating plant closings and job loss in the United States. Workers in operations that do remain on U.S. soil will be forced to negotiate diminished wage and benefit packages under constant threat of production moving to other countries. This is already happening as a result of the NAFTA treaty.

If this type of treaty isn't a threat to democratic control of our own lives and our ability to safeguard the earth on which we depend for life itself, what is? How is it that we can heroically respond to natural disasters like floods and earthquakes, but sit by while the power to shape our lives and protect our children's future is being taken away from us in the guise of an investment agreement? Talk about white collar crime!

MAI's Rapacious Offspring

MAI-type tenets are now being proposed in several other trade agreements under negotiation: the Free Trade of the Americas (FTAA) between North and South America, the Asia Pacific Economic Cooperation Forum (APEC), which includes most Pacific Rim countries, and the proposed TransAtlantic Economic Partnership (TEP), which links economies in the United States and Europe. Even more insidious, the International Mone-tary Fund (IMF) is campaigning to amend its Articles of Association so it can demand that recipients of its aid make their countries even more favorable to foreign investments.

Order a copy of the clear and concise booklet A Citizen's Guide to the World Trade Organization.

Send $2 to Apex Press, Suite 3C, 777 UN Plaza, New York, NY 10017; or call 914-271-6500.

Citizens Unite!

It doesn't have to be this way. We do have options. We can demand that the WTO do an evaluation of the wage, health, and environmental effects of the 1994 GATT treaty before expanding into new areas and codifying the MAI in a "Millenium Round." We can create investment agreements that hold corporations and investors to baseline standards of environmental and public health protection, fair wages, and safe working conditions. We can put into place mechanisms that allow local and indigenous peoples to keep investors accountable. We can make sure there is no provision for corporations to be granted monetary compensation for the outrageous notion of "loss of future profits." We can make sure that we hold onto our democratic right to write laws that promote the common good. We can set the process up so that each draft of new agreements is publicly released in a timely manner, with open meetings scheduled in all countries.

Trade agreements are not irrevocable—especially those that haven't yet been voted on by Congress! We have not only the opportunity but also the imperative to claim what we want for our lives now and to create a new economic destiny. Where to start?

Write, fax, or e-mail your elected officials, both local and national. Write letters to the editor. Get an article in your church bulletin or professional newsletter. Call talk radio.

> *Join* Public Citizen's *"Trade Watch" to be kept abreast of the latest developments in trade agreements. Call 202-546-4996 or look up their web site:* www.tradewatch.org.

Is this a crucible? Yes. Is it a reason to feel hopeless? No. Let me tell a story.

A Zen master was out walking with his student. The student saw a fox chasing a rabbit and said to his master "I wonder which one will win?" His master replied "According to an old fable, the rabbit will get away from the fox." The student said "Surely not, the fox is faster." "But the rabbit will still escape," insisted the Zen master. "Why are you so certain?" asked the student. "Because the fox is running for his dinner, but the rabbit is running for his life."

Questions For Reflection

- What feelings come up when you think about trade agreements being negotiated by corporations and passed by our political leaders that strongly favor investors, stockholders, and CEOs over working people and the environment?

- Where do you see your life in the confluence of local and global forces?

Chapter 8

Time: A Nonrenewable Resource

Suzanne Stoddard

why we aren't finding time to live

Time is the substance I am made of. Time is a river that sweeps me along, but I am the river; it is a tiger that rips me apart, but I am the tiger; it is a fire that consumes me, but I am the fire.
—Jorge Luis Borges

Let's not look elsewhere for the basic reasons for a sense of having too little time: our desires have increased much more quickly than the available time.
—Jean-Louis Servan-Schreiber, *The Art of Time*

Think of many things—do one.
—Portuguese saying

There were times when I could not afford to sacrifice the bloom of the present moment to any work.... Sometimes, in a summer morning ... I sat in my sunny doorstep from sunrise till noon, rapt in a revery, amidst the pines and hickories and sumacs, in undisturbed solitude and stillness, while the birds sang around or flitted noiseless through the house, until by the sun falling in at my west window, or the noise of some traveller's wagon on the distant highway, I was reminded of the lapse of time. I grew in those seasons like corn in the night, and they were far better than any work of the hands would have been.

—H.D. Thoreau

Remember the good old days when there was *time* in life? When conversations weren't always rushed, people strolled instead of jogged, playing a game could take hours, and no one minded? Today, few people in our culture besides the very young and the retired are exempt from the tyranny of time pressures. We come home late at night, tired from working overtime and hassling with traffic, and still have dinner to fix, dishes to wash, homework to help with, bills to pay, and laundry to do. Even having to return a friend's phone call can be taxing when it's one more task in an over-busy day. An activity that should be pleasant can feel like *one morecommitment.*

Ticking Clocks or Time Bombs?

Who or what is responsible for this sense that we're living in a revved-up universe?

One reason our time is congested is that we have so many more people competing for a slice of us than our ancestors had. Today's sophisticated communication technologies allow everyone we know—and many we don't—to trespass upon our time. Even if we opt not to respond to them all, we still have to wade through all the e-mail and phone messages.

The ethic of progress in our economy is "produce and consume." This gets to the heart of explaining why time has gone out of life. So many of the products we buy, rather than saving us time, *take* time to use, maintain, and repair. The advertising messages we think we're impervious to make our desires grow, while no one has figured out how to expand the 24-hour day.

The highly competitive nature of modern Western life—the unavoidable consequence of a produce and consume culture—is the most important reason that time has come to feel like the enemy. To keep a job today often requires the willingness to endure a long commute, work more than eight hours, network with colleagues, schmooze with clients, buy new office clothes, and learn new business and computer skills to stay current. All these activities use up our time, leaving only slivers for ourselves and our families. (We realize that those who are blessed with jobs they love may find equal or greater satisfaction at work than at home.)

The madness of speeded-up time is taking a huge toll. Too often we need to be in three places at once and feel guilty that we're not. This impacts the quality of the present moment. Stress-related illnesses for men and women are at an all-time high. Not long before French writer Jean-Louis Bory took his own life, he was quoted as saying about his digital watch "This watch, you see, is my death."

Indigenous cultures have a different relationship with time. Their labors, religious ceremonies, dances, and feasts are interwoven with the rhythm of the seasons and the heartbeat of the earth. No wonder they find Western time patterns disorienting, unnatural, and destructive. Having had their ancient ways of living and working decimated, Native Americans have faced the agonizing choice between confinement on reservations or struggling to integrate into urban life. In both choices, the cultural discordance has produced drastic increases in substance abuse, suicide, divorce, and juvenile crime rates.

What We Lose When We Go Too Fast

At our modern breakneck pace, too much is sacrificed. We lose touch with what makes us feel good, with what gives joy. We forget that we used to delight in our garden or in a walk in the hills. We fall out of step with our natural rhythms. Speed puts us out of sync with our biology and out of harmony with the earth. After all, life was never meant to be a frenzied sprint from one activity to the next.

The sheer quantity of stimuli assaulting us makes it easy to lose our sense of purpose. If we're unable to decide what to focus on, we can easily become depressed or paralyzed and incapable of immersing ourselves in anything. Then, too, our multi-tiered obligations make us feel that we can never relax, that something else always remains to be done.

Paring down our wants and needs isn't easy in this culture, which pushes us to seek material goods and rewards. Although no one likes to be made a fool of, we are duped every time we are persuaded to buy something we don't need. Millions of Americans purchase expensive, worthless products from the TV shopping channels. The former husband of a friend of mine is so addicted to such purchases that he has incurred thousands of dollars debt for useless items that now fill an entire room. Even disposing of things no longer wanted takes time. By the simple act of turning off the TV, we *could* affirm that we value our time, money, and the beckoning of our heart. We could explore our own inner promptings instead.

Our own lives and dreams are at least as compelling as those of the celebrities du jour, but we lack visible contemporary role models for slowing down, calming down, and looking inward to discover our deeper desires. While enlightened individuals who could be our guides on this journey do exist, we rarely hear about them in the mainstream media.

How It Feels Not To Have Time

Although feelings are often devalued in this analytical, hi-tech culture, they are the key to what's really happening inside us, why our life is working or not, and what level of satisfaction and contentment we are experiencing.

How do we feel when there's no time for enjoying music, partaking of leisurely meals, meandering through magazines, savoring intimate conversations with friends, or just puttering around? How do we feel about ourselves? How do we feel about life?

The bumper sticker "Are we having fun yet?" says it all. There is no time for fun anymore, because our lives are shredded into ever smaller strips in which to do more and more activities. Books like *The Overworked American* by Juliet Schor attest to the seriousness of this growing phenomenon. Most things of value take time. Yeast rising for bread takes time. Listening attentively to our children takes time. Figuring out our own underlying feelings takes time.

Ask yourself: When do I truly come alive? If you haven't really relaxed for a long time, you have probably forgotten the sensory delight of sunrise over the mountains, smacking a tennis ball, cuddling a kitten, fondling several fabrics before finding the right one for a new sewing project, wending your way through a mystery novel, taking a nap, or photographing the dew on morning grass. Indulge yourself by setting aside a minimum of two hours a week to do what you love. *Recreation* is a beautiful word, because what we are literally called to do in free time is re-create ourselves. Try turning off TV for a day or a week. Revel in the time that opens up!

Give yourself an entire day with no commitments. If this is impossible, carve out at least two hours. Sit by a running creek. Go walking in a place you love. Fool around with a craft project or paint a picture.

The more oppressed we are by time, the less kind we are to ourselves and others and the less able we are to rise above life's difficulties. But who says we have to submit to our culture's prevailing view of time? We can refuse to keep buying into a system that's out of balance by making more time for what stirs and renews us.

Drop at least one item that you don't enjoy from your daily to-do list if it doesn't absolutely have to get done.

The Richness of Unmeasured Time: Two Personal Stories

I surely would have missed the best adventures of my life if I'd been in a hurry. The quality of experience and interaction has been richer when I've taken time to fully engage in the moment and slow down to my natural setpoints. When we put time constraints aside, unexpected joy can find us.

This is why I love road trips. Driving, rather than flying, lets me ease into my free time gradually and uncoil from the whirlwind of work and preparation that always precede a trip.

In the summer of 1995, I took a driving vacation to Wyoming with Montana, the 13-year-old daughter of my best friend. With the windows wide open, we sang loudly with the cowboy songs on every country music station we could find. I, who had never even liked country music, found myself not merely enjoying but enthralled by the maudlin words, the universal love-and-loss themes, and the repetitive choruses.

Opening up time allows us to view the world through fresh eyes and makes experimentation more appealing. Conversely, when time is limited we tend to focus on activities proven to give satisfaction and close off the possibility of surprise.

Once Montana and I arrived at the dude ranch, we had nothing but time—to eat, ride horses, sleep, read, talk, help with the chores if we wanted, or just laze around. On one particular day, I opted not to go out for the morning ride but to help the two ranch hands, Randy and Ted, mend fences. The morning was sunny with an intense blue sky. The air was light and fragrant. My sturdy companions deftly measured, unrolled, clipped, and tied the barbed wire to the thick wooden fence posts. A couple of hours into the work, Randy noticed a prickly pear that had mysteriously fallen near us (there being no cactus in sight).

The gift of this exquisitely formed, perfectly ripe fruit was a numinous treasure. Randy gently prodded the light green skin open with his knife to reveal golden pink flesh on the inside. Slicing equal sections for Ted, himself, and me, this Choctaw Indian leaned back on his haunches, seemingly in another time, as though honoring his ancient ties to the natural world. Randy's quiet sense of awe affected me and Ted as well. We ate our sweet delicacy in worshipful silence, in a measureless moment of unanticipated communion.

Although the week was rich in starry nights, home-baked pies, languorous naps, and horseback fun, of all the things that happened it was the prickly pear morning in the pink granite mountains that I remember most. A forgotten ember deep within me had been stirred, and I realized that when people talk about spiritually transformative events in their lives, they must mean moments like mine, of unexpected wonder and quiet joy. I'd been offered a glittering window on a larger way to live-a way that was simpler, deeper, brighter, and shimmering with possibilities.

The Wyoming adventure made me realize how estranged I had become from myself. Rushing from one deadline to the next in my corporate job, I had temporarily lost touch with my natural rhythms. How many times in the last few years had I overlooked chances to savor a seemingly small miracle of nature or human connectedness simply because I hadn't take the time? Wyoming gave me a new feeling of appreciation for my life and for the moment-by-moment potential for magic.

Transcendental time again opened up to me following a conference I had organized for Pacific Gas and Electric. It had taken weeks to plan, including many bleary-eyed late nights in front of the computer screen. Finally, the three-day conference materialized in Santa Cruz, ending on a Friday afternoon. It had been very successful, but I was too exhausted to fully enjoy the feeling. The last thing I wanted was to fight rush-hour traffic back to the Bay Area. I tried to get a complimentary extension on my room, but found that it was already booked. So I bit the bullet and paid for a room at a different motel on the beach. I fell into bed mid-afternoon and slept like a dead woman.

Hours later, I awoke to shafts of vibrant light dancing across my room. Refreshed from my sleep and intensely curious, I ran outside and saw flaming splashes of crimson, orange, and pink suffusing the sky. Frisky puppies, windblown beachcombers, and laughing children were haloed in a celestial glow.

I felt like an honored guest at a primeval banquet.

Twilight brought another kind of perfection, as the tide-washed, pearly gray sand near the water luminously mirrored the navy blue sky. Quiet voices, soft barking, and seagull cries completed the enchantment.

I had merged with the cosmos in a moment beyond time, like a fully awakened being wholly in tune with humming consciousness.

Would I have had the privilege of such ecstasy if I had been rushing? Not likely.

Questions For Reflection

- How do you feel inside when you are rushing all day? Could some of the rushing be avoided?
- When do you feel the most free? What is your sense of time in these moments?
- What eats into your time each day? How could you eliminate or reduce activities that gnaw away at your precious time?

Part II

Fresh Choices: Saying Yes! to a More Expansive Life

We are at a critical choice point about the future—our future. The corporate way of life has so revved up the pace of our activities that most of us feel stressed and anxious much of the time. Even when we are not under a crushing deadline, we know that another big project to be completed in a heart-stopping timeframe is just around the corner. The time between major work tasks is far too short these days to savor what we've accomplished. And the spillover of job stress contaminates our off time with our families, turning leisure hour into arsenic hour.

We have become too accustomed to "rolling with the punches," however stomach-churning they may be. We forget that we don't have to accept the way things are, that we are choice-making creatures. When the carpet is being pulled out from under our feet, it could be a signal from the universe that it's time for a big step. What makes us afraid is that we don't trust ourselves. Choosing to make a big change can seem scary, even foolish, no matter that our familiar routines are not working. Using our intuition we can make very good choices, but most of us have not learned to develop this part of ourselves. We live in a culture that doesn't encourage slowing down, going inside, or finding balance.

Why do rivers call to us so deeply? Because they connect us to our yearning to be in the flow of life. They remind us of our dreams and visions. They remind us of our innate joy and freedom. When we look at Nature, she is ever changing, yet always at peace. We can use this as an exquisite metaphor for our lives: being vibrantly engaged in the outer world, while staying balanced and focused in our core.

For both the inner and outer journey, courage is key. It takes courage to look deep within, and it takes courage to tackle an issue in the economic, political, environmental, or service sphere that we may never have confronted. Finding our courage does not mean conquering our fear. It only means that we no longer let fear stop us. We do make mistakes as we test our wings, but every experience reveals to us another aspect of what we need to know. As we create safe spaces for experimenting with new ideas and behaviors, life becomes a much grander adventure. Interesting new people who share our values suddenly join us for parts of our journey because something inside us has expanded to include them.

Part II explores the personal situations that are grist for the mill of empowerment. It plays with the differing perceptions that can become flashpoints or transformations. In situations of miscommunication and disagreement, the heart is a profound mirror. It reveals to us our deepest truth, which, if we act on it, will move us toward wholeness. The clarity and strength gained from these experiences powerfully fuels effective action in the outer world: creating new work, reclaiming the global economic commons, safeguarding the planet, and creating nurturing bonds of community.

Chapter 9

Change and the Comfort Zone

Ellen Schwartz

embracing risks that have been foisted upon us by life

The greatest mistake you can make in life is to be continually fearing you will make one.
—Elbert Hubbard

It is not the mountain and the river that defeat us, but how we think about the mountain and the river.
—Vietnamese saying

Change is one of the most unpopular concepts on the planet.
—Caroline Myss, *Anatomy of the Spirit*

Change is the word that holds the most hope and the most fear.

Even when our lives are full of pain, even when we're aware of a constant low-lying anxiety, it's still hard to welcome change. Leaving that which we know takes courage.

For good reasons we are afraid of change. For good reasons we are afraid today. We fear losing our job, and then our home. We fear those who live around us; the nightly news shows us graphically what they who are different from us do, how they rob and rape and murder. Perhaps fear is the predominant characteristic of our times. Perhaps history will remember us as the age of relentless technology, frenetic activity, and fear.

I remember reading about Pavlovian dogs in college. They were given food each time a bell rang, so they were gradually conditioned to salivate simply at the sound of a bell. I remember how repulsed I was at the suggestion my professor made that we, too, are conditioned. Yet it's clear to me now that we *are* conditioned—especially to comfort. The advertisers who dominate our visual space, inside and outside our homes, have conditioned us to believe that comfort is the American Way. Alexis de Tocqueville noted more than a century ago that Americans feel their every whim and fancy should be gratified. It took Madison Avenue advertisers and the technology of television to perfect the idea that instant gratification is our birthright.

Yet the desire for comfort is the greatest enemy to our growth. Under the influence of comfort, we rarely leave our known environments, our familiar pain. The changes we so fear and rail against in the beginning, however, often turn out to be the greatest gifts. Such was the case for me when I was sixteen.

At that time, my asthma was so bad that I was spending two to four weeks a year in the hospital. My doctor said Detroit was too humid and pollen-ridden; he suggested a drier climate like Denver, San Diego, or Tucson. And so it was, on a family vacation in Arizona, I found myself choking back tears, waving goodbye to my family from the doorstep of a small Catholic school in Tucson. I had thought we were going on a dream vacation to see the Grand Canyon and the vast Western spaces. My dad, fearing asthma would turn to emphysema, had intended from the start to find a boarding school for me.

The depths of my terror and aloneness are hard to describe. Having asthma had meant a rather solitary childhood. I couldn't run and play with the kids at recess; after school, I couldn't ride my bike the couple of miles to my best friend's house. By the time I was entering senior year, I was quite attached to my little world of family, a few close friends, and the books I loved.

Living on the lower rungs of the middle class, my parents couldn't afford to leave their jobs on the chance that a climate change would improve my health. There was really no choice but to leave me in

Tucson by myself and see if it would work. Every night from Labor Day until I got to fly home for Christmas, I cried myself to sleep. While my heart ached, I think my dad's heart was broken clean in two. Every morning he got up at 5:30 so he could write me a letter before work. Some of the envelopes were plump with photographs he'd taken of the resplendent autumn reds and golds we both loved. At Halloween, he sent me a big box of fragrant Michigan apples, my favorite red and black licorices, and carmellos. And when I came home for Christmas, I found that this simple carpenter had gone to a fabric store, bought a bolt of red cloth, and stretched it from the sidewalk to the front door, to "roll out the red carpet" for me.

My asthma did improve remarkably. Even so, Dad gave me the choice of staying home after Christmas or finishing the year in Tucson. I had so missed my known world that up until a couple of days before my flight, I was determined to come home to stay. Not that my home was ideal. The tension from my parents' arguments was palpable. And it was wonderful to draw clear, deep breaths in Tucson. Still, Detroit was familiar, it was what I knew.

One evening just before the Christmas holidays my English teacher, Sister Rachel, came to talk to me. She knew how difficult the past months had been. I had written about them in my journal, as part of her class assignments. She told me about another girl she knew who had left home for college. This girl had been so miserable that she returned to her family after only one semester, and now, fifteen years later, she was still afraid to leave the comfort and security of her family though she longed to see the world. Sister Rachel told me she feared that, once I went home, it would be very hard for me to venture out again. I thought about that story throughout the break, and got back on a Tucson-bound plane on January 2.

Up to that point I had been very shy, but now I decided I might as well make the best of things. I joined the yearbook and art clubs and volunteered for every committee. I threw myself heart and soul into getting to know people and being out in the world. I found I liked it very, very much! I discovered that people whom I had found so intimidating were really wonderful human beings. I discovered that I loved the thrill of working in a group and making a project coalesce. I discovered strength and creativity in myself that I had had no inkling of when I was living in Detroit.

I changed from an introvert to an extrovert that year.

Most of all, I realized I would never be afraid to put myself into an unknown situation again. I went to a college in Denver where initially I knew no one. My junior year, I transferred to a university in the San Francisco Bay Area, where I knew only one person. I talked my way into a lot of jobs where I didn't really have the right experience, because I had enthusiasm and the will to learn. In my thirties, I went to the Soviet Union, Reagan's "Evil Empire," and returned home determined to bring two warm peoples together in citizen exchanges. Then in 1994, I ran for U.S. Congress against all odds. Though I did not win, in my first bid for public office I got the most votes of any person running against an incumbent on the West Coast that year. I know I would never have dared such a feat had I not been left standing on that Tucson doorstep when I was sixteen.

Try something new this week that takes you out of your comfort zone. Write a note to yourself about the results. Do it at least once a week.

If we are to allow what is within us to emerge, if we are to let our spirits soar, we must give up our desire to be on familiar ground at all times. Our desire for comfort dulls the still small voice within that beckons us to try something new, to reach out to someone else, to be a part of something beyond ourselves.

How do we ease away from the seduction of comfort? Sometimes, it is Life which ejects us from our comfort zone. We still, however, have the option of going with the flow or bitterly fighting it. Learning by crisis is one way. Learning from inner promptings is another. In our everyday lives, we can begin by creating a quiet space. Away from the noisy stimuli of television, radios, CDs, and computers, thoughts and feelings begin to bubble to the surface. Out of the rush and away from the hurry, we begin to both feel the pain we have been avoiding and the passion of our soul's calling. Is it easy? Rarely. Is it worth it, living in *yes?* Absolutely!

> *What is your biggest dream for your life? What stops you from reaching toward it? What are one or two small steps you could take now to move yourself forward?*

Questions For Reflection

- What potential change are you most afraid of? What would you lose? What might you gain? How much energy do you expend dreading this change? Do you think the low-lying anxiety about having to face it takes a toll on your physical or mental energy?
- How much does comfort mean to you? What are the baseline comforts you "can't live without"? Do you ever feel that your attachment to comfort stops you from exploring something new?
- What is your usual reaction to change? Do you want to shut the door immediately, or do you start mulling over possibilities?
- What helps you think things through and feel your deepest feelings? Journaling? Walking by yourself

in nature? Meditating? Are you willing to create time and space for these on a regular basis?

Chapter 10

What's an Inner Life and Who Needs It?

Ellen Schwartz

trading fear for trust and planting seeds of loving-kindness

Renew thyself completely each day; do it again, and again, and forever again.
—H.D. Thoreau

My religion is kindness.
—Dalai Lama

Religion and Dogma dominated my childhood and early adult-hood. We were Catholics, the "one, true faith." The other Christian sects were unfortunately misguided, and as for those poor savages in the underdeveloped world, we saved our candy money during Lent to pay for missionaries who would "save the pagan babies."

In my late teens and early twenties, when I would come home from college for the summer, I took to driving to the cemetery and parking beneath the old sprawling trees during the hour that I was supposed to be at Mass. Usually I wrote in my journal. I did not want to take on the battle with my mother, who followed the letter of the Catholic law.

Something in me led me to look at the beliefs and practices of other peoples and other times. As I wondered how all these other ways could be "wrong" and "condemned," Pope John XXIII was convening ecumenical talks, respectfully meeting with leaders of other faiths.

The more I explored other traditions, the more I felt that each held an exquisite piece of the universe's mystery. Indigenous peoples bless the sacredness of the earth and see all of creation as an interconnected circle. Hindus bring human sensuality into the realm of the sacred. Buddhists bring to light the nature of "illusion": those situations we spend so much energy raging against and seeking revenge for that are actually settings for our own courage, endurance, equanimity, or compassion to emerge. Sufis focus on the harmony of love, joy, and beauty, where the essential oneness of all Being is experienced.

Spirituality has come to mean for me an intention—to grow, to serve, to delve into mystery. Gratitude, which I have found to suffuse all the traditions, permeates my journey: gratitude for the unconditional love of my father, for the beauty of this earth, and even for the people who present the challenges that test my humor and tax my patience. I have come to believe that God is no less with me in times of chaos than in times of happiness. I have come to believe that trusting is one of the toughest and most powerful of life's lessons.

I have spent a lot of time microplanning my life and fretting over details—in other words, trying to circumvent the necessity for trust! Something shifted in me when I did the previously unthinkable and said yes to running for Congress. I was in uncharted territory, not operating on the usual assumptions and expectations. I began to notice that, if I needed to meet someone, within a few days I would run into them. If we needed furniture or supplies at the campaign headquarters, they would turn up just in time. I began to understand what was meant by "The path has been prepared before you." When the campaign ended, although I was surprised that I didn't win, I didn't feel beaten or defeated. I had a strange sense that all was well.

As I look back, I realize that I was completely absorbed *in the present moment.* I had heard that phrase decades earlier, but tossed it aside as trivial or even nonsensical. After the campaign was over and I had the chance once again to pursue spiritual readings, I began to see that living in the present moment is a fundamental, pulsating truth in all the great religious traditions. To be grounded where we are and aware of the feelings and emotions washing over us, while being open to emerging possibilities, takes attentiveness. Then, too, we need to duel with the ghosts from the past that rush to label and control fresh situations.

Sometimes we know exactly why we're feeling as we do, but often our rage, confusion, or hurt comes from a place so primal that we don't immediately understand what we're experiencing. Journals are great friends when we can extract ourselves from the maelstrom to write whatever comes. Nonlinear practices can help ease us into a greater state of clarity. These can be: sitting quietly, breathing slowly, watching the sunset, or joining in the Dances of Universal Peace (doing simple circular movements while singing phrases from all the world's great spiritual traditions). Chanting, whether ancient mantras or contemporary phrases, allows a deep resonance field to take us beyond our twentieth-century stresses and worries. T'ai chi—slow, graceful gestures imbued with timeless meaning—is calming, relaxing, and balancing.

Thich Nhat Hahn offers another possibility when the pain or confusion is long-lasting: plant a seed of loving, kind, or compassionate action. Water only that seed with your thoughts and energies. The seedling will burst upwards into the sunshine; the negativity, shaded out, will gradually wither and fade away.

Thich Nhat Hahn tells of a man who attended a retreat he gave for Vietnam veterans. This man's unit had been part of the My Lai massacre. He and several other soldiers had stormed a schoolhouse, gunning many children to death. In all the years since, this man could not bear to be in a room with children. Thich Nhat Hahn said "You cannot bring those children back to life. But there are many children now who need your help. Some poor children need immunizations. Some need help learning to read. Go and help these children who are here, now." Years later, the man contacted Thich Nhat Hahn to say that in that action he had finally found relief.

Relentlessly examining our feelings, reactions, attitudes, and agendas is a necessity for spiritual growth. Once a day is a good start. Looking nonjudgmentally at those times when we fall short gives us the chance to frame an apology or set a fresh path. Being a recovering perfectionist, I still find it much easier to forgive other peoples' shortcomings than my own. But, as the years go on, I am more and more able to say "Oh, well."

One of my favorite Tibetan Buddhist teachers, June Rosenberg, helped me release some of my judgments of myself and others. She suggested that for one week the class focus our awareness on judgmental thoughts, and each time we found that one had arisen, say to it "Blah, blah, blah!" When repeating this, it is nearly impossible not to laugh at our own pomposity.

For another week's assignment, we examined our masks. Were we wearing our Self-Righteous mask, our Victim mask, our Corporate mask, our Pillar of the Community mask, our Rebel mask, or yet another one? Masks block others, and often ourselves, from knowing our true thoughts and feelings. They keep us from living in the present moment.

Even if you are certain that you don't wear masks, try going through a whole day with responses and reactions that are utterly authentic and from the heart. Notice how you feel and how others respond to you.

Can you imagine a world where you could be fully honest all the time? When people asked you how you felt, you answered straightforwardly, without anger, blame, or "I told you so"? When your boss asked you what you thought of his latest idea, you gave him your true opinion of its strengths and shortcomings? Can you imagine the toll on our bodies and spirits from all the times we have *not* spoken our own truth? Or the toll on others when we've spoken it without humility, making it very hard for them to hear us?

Caroline Myss *(Anatomy of the Spirit)* puts it this way: "We are not humans on a spiritual journey, but spirits on a human journey!" Not only does our smallest action or omission play into the workings of the universe, but each thought and intention does also. To be awake and attentive in every moment is no small task, but the more commitment we make to stay conscious, the more carefree and joyful we become.

What makes joy permeate your soul? Being at the ocean or mountains? Singing? Dancing just with you and the music? Watching the star-studded night sky? Make an appointment with joy this week!

Questions For Reflection

- What renews you? What takes you to a deep space beyond time where you feel Oneness, connected to and in the flow with something much bigger?
- Think of a time when you marveled at the coincidence of something. Call up another time when you were delighted at the serendipity of an occurrence. Remember a spontaneous event where you were flooded with gratitude. Do these things seem part of a seamless whole?
- If you attune to a particular religious tradition, what in that belief system resonates with who you truly are? At what points do you diverge, listening instead to your own heart and mind?

Chapter 11

Flashpoints

Ellen Schwartz

how our stresses play out in the family crucible, damaging those we love most

Make it thy business to know thyself, which is the most difficult lesson in the world.
—Cervantes

Man is the only animal that blushes—or needs to.
—Mark Twain

We have committed the Golden Rule to memory; let us now commit it to life.
—Edwin Markham

The tension inherent in keeping the mega-economy running holds our lives in a tenuous position with those we love most. We try so hard to make everything work, but even so, fear lurks just around the corner—the fear that we could be downsized and "lose it all" tomorrow. Living on the razor's edge, even small problems can trigger major eruptions.

For a long time I have been trying to learn on a cellular level that the immediate gibe always causes more problems than it solves. The Dalai Lama says "Our first instinct is always to retort, to react, and sometimes to avenge ourselves, which brings on nothing but more suffering."

Having been married for twenty-three years and been a parent for twenty-two, my life is strewn with experiences of having pounced on people in instant judgment. Even while I was raving, somewhere inside I knew that criticizing and pontificating were having no impact. The problems were just getting bigger, and now I had defensiveness to deal with along with the original difficulty.

When I step back from the situation (usually after I've made a mess of things and then have had time alone to stew about them), I generally become aware that when I get angry at someone else it's because I'm disappointed in myself. Either I haven't accomplished all the things I needed to do, or I've eaten foods I know drain my energy, or I haven't been getting enough exercise to stimulate and balance the mind/body loop. When I'm at my physical best and feeling reasonably good about myself, I can handle an unexpected or troubling situation by examining the problem itself and not make matters worse by jumping to conclusions and letting others have it.

I am often surprised that, when a conflict occurs and I take the time to check out all the angles with the other people who were involved, I find I wasn't totally right, nor was my instant solution the best. Taking time is so important. There is no way to understand a situation without deliberately stopping and taking the time to listen. Only when everyone gets to say what the "facts" were from their point of view, share their feelings about the situation, state their intentions, and acknowledge "ghosts from the past" that seem to be overshadowing the event, can clarity be possible. A solution dictated by the person with the most power has little chance of enduring past sunset. But when everyone brainstorms ideas and has a say in the process, there is a good chance of a long-standing agreement.

What can you do to take care of yourself so you're more in a state of balance? What hungers can you feed, what needs can you tend to, what dreams can you nurture? Think of one. Do it now.

Gandhi observed that while using force, either verbal or physical, seems to be the most direct and powerful way to get change, it is really no solution.

It only breeds resentment and revenge. The longer and seemingly slower route of deep listening, compassionate persuasion, and nonviolent action is the only resolution that holds.

I had the good fortune many years ago to move onto a street where a wonderful older couple live. George and Elsa are retired teachers. They embody both calmness and a mirthful zest for life. I remember once when we had them for dinner, Elsa related an incident which, had it happened to me, would have made me boiling mad. At the end of the story, Elsa said "I was most disappointed."

I was flabbergasted! How could she have had such a different reaction than I would have had? But as I began to think about Elsa's story, I realized that her response had been the wiser. My reaction would have only fanned the flames. I began to realize that anger is really a secondary emotion; what comes first is either hurt or frustration. Elsa had been frustrated, but her approach of stepping back and observing the situation ultimately let her see the bigger picture and take an action that made sense.

I am still a novice at handling my own overreactiveness. What works best for me is the age-old advice: Don't talk. Take a deep breath. Walk away. Discuss the situation only after time has elapsed and emotions have cooled. Two or three slow breaths would be optimal, but managing just one without blurting out a judgmental comment is a near miracle for me in the heat of conflict. Walking away when I want to prove how right I am takes enormous presence of mind. Waiting to discuss the situation stretches my tiny reserve of patience to the max. But I can tell you, this works.

In the West, we have been schooled in the theory of reality that "only what is visible and measurable is real": the chair is real, the car is real, and only the words that pass your lips have an impact. However, in trying to live with other human beings in such a way that conflicts turn into little jewels of growth, I have found that what is unsaid and unseen has great power. Intentions take over a situation and are perceived unconsciously by those involved.

In the early stages of writing this book, Suzanne and I were given the wonderful opportunity to stay at a friend's vacation home near the ocean. Our ten days were utterly magical. We savored the pure delight of being able to eat, sleep, and drink the ideas we were pondering; the play of pounding rain and bright winter sun on coastal vegetation; the vigorous walks by the ocean at high noon and the luscious, long sunsets.

I thought I was in a state where nothing could disturb my equanimity upon returning home. But just walking through the door into the family room triggered my emotions. We have an electric heater that we turn on only when we're using the room because it makes our utility bill jump enormously. I was the first one home and the family room was *hot.* I guessed that the heater had been on day and night while I was gone, and visions of the extra hundred dollars on our electric bill took over my consciousness.

I vowed to myself that I wouldn't let a trace of anger tinge my voice when I brought up the subject. I'm the kind of person who wants to talk about things right away, get everything out, and be done with it. This time, however, I would make the supreme effort not to bring it up right away; rather, I'd give big hugs and ask how everyone was doing. My husband and daughter arrived shortly, and the first five minutes were great. Then I asked, manufacturing the most neutral tone I could muster, "Has the heater been on for ten days?" The thunderclouds rolled in over my husband's face, and the night was destroyed. "I've done everything for the kids for two weeks, and you come down on me for the *heat?"*

I had my ducks in a row. First, he had exaggerated—it was ten days, not two weeks. Second, I had taken extra time to shop before I left to make sure the house was stocked with easy foods and goodies. Third, big deal taking care of the kids—my son was twenty, and my eleven-year-old daughter was more capable and resourceful than most adults. Fourth, it looked as though he was guilty as charged: the heat must have been on around the clock the whole time. And fifth, I had deliberately used a "nice" tone of voice rather than an accusatory one.

Of course, I didn't get a chance to lay out my self-righteous arguments. My husband was much too mad to listen to "reason." Saturday evening the tension was still palpable, so I suggested we drive to a place overlooking the Bay Area lights to talk. My husband said "There's nothing to talk about" but agreed to go. My opening comment of "I'm sorry I brought up the heat so soon, but you've got a part in this too" didn't get the conversation very far. Michael yelled "You're completely to blame for this. It was the coldest January we've ever had. The TV room never got warm enough just turning the heat on after work." "Why didn't you explain that last night?" I asked. He jerked the car around, and again the night was a washout.

It took me another day before I could admit to Michael that I was furious under the "calm" facade. Those angry, judgmental vibes underlying our actual words were the real determining factors in what transpired. What was unseen and unspoken turned what should have been a five-minute conversation into a lost weekend of bad feelings.

Sometimes I can catch myself sooner and bring the unstated into the open. Sometimes I can more easily identify my auto-responses, my masks, my underlying feelings. And as I grow older, I become slightly more forgiving of myself when I rant and rave needlessly.

Try this experiment when a disagreement comes up. Remove yourself from the situation and write brief notes about what surfaces for you. When you come together to talk about what happened, set the notes aside. Listen to the other person with the full attention of your heart and mind. When it's your turn to talk, incorporate the other's genuine concerns and pain into your own perspective.

Questions For Reflection

- Think of a recent experience where you "lost it." What was the apparent trigger? What emotions or fears underlay it? Were "ghosts from the past" (recurring similar instances) putting your reactions on auto-pilot?
- Name three topics that evoke charged reactions from you. Do judgments—of yourself and other people—fuel these reactions?
- What happens when you try to talk through a situation in the heat of the conflict? Does it get more or less complicated? Do you feel you are able to listen deeply to the other person without preparing your own counter-arguments while they're talking?

Chapter 12

Nurturing What Is Precious

Ellen Schwartz

finding new ways to communicate and connect with our loved ones

Children have never been very good at listening to their elders, but they have never failed to imitate them.
—James Baldwin

Your children are not your children. They are the sons and daughters of Life's longing for itself. They come through you but not from you, and though they are with you yet they belong not to you. You may strive to be like them, but seek not to make them like you. For life goes not backward nor tarries with yesterday. You are the bows from which your children as living arrows are sent forth...
—Kahlil Gibran, *The Prophet*

A Mom's advice: Trust your intuition. Be kind. Eat good food. Tell the truth.
—Gaye Frisk Lub (artist)

The little kindnesses and courtesies are so important. Small discourtesies, little unkindnesses, little forms of disrespect make large withdrawals. In relationships, the little things are the big things.
—Stephen R. Covey, *The 7 Habits of Highly Effective People*

There is a Buddhist tale about an ascetic monk who lived in the mountains. The King had asked him repeatedly to come down to his palace to teach him and his courtiers. Finally the monk agreed.

All was going well as the weeks progressed. The King was happy to be receiving the monk's teachings. Then one day, the palace fell into an uproar. The Queen's favorite jewels were gone! A great search was made, but the gems were nowhere to be found.

The next day, the monk appeared before the King's throne. He placed the jewels at the King's feet. "You?" said the shocked King. "I never suspected you!"

The monk replied "Here in your palace I have been eating the food prepared by your cook. He is a man whose heart is filled with greed. As he cooks, his thoughts and feelings enter the food. I was acting under the influence of the greed I had taken in. I am sorry. I must return to my mountain."

Clearly, what we think and feel has a powerful effect upon those around us. Our *intentions* have just as big an impact on our children as our words. To be sure, we want to choose our words carefully, but our thoughts and personal agendas, indeed, our very presence, also hold great potency. How we live our lives reverberates in our children's souls.

As Gandhi said, "You must *be* the change you want to see in the world." This means that what we think, feel, say, and do must be one and the same. Before we had children, it didn't matter quite so much if we occasionally used foul language, or were terrible procrastinators, overindulged in food or drink, or lived in chaos. Now it matters immensely. We are the people our children model.

Living each moment in consciousness is a very large task. But the more we state an intention to be aware, the more often we remember. Turbulent emotional states seem to have a way of creeping up on families. There isn't a mother or father who hasn't at one time or another wanted to shake the cookies out of their little darling. Taking time out, as soon as one person becomes conscious of the anger buildup, helps a lot.

Of course, everybody's feelings and attitudes add to the boiling pot. When interrupting a volatile situation to try to get some clarity, it helps to come together in a large space where everyone can sit in a circle. Establishing the ground rule "no interrupting," each person takes his or her turn answering the levels of questions: What are the facts of this situation according to your point of view? What are your feelings now? What were your intentions? What assumptions or judgments do you think are impinging on what happened? What are your best ideas for possible solutions? Once all this has spilled out, answers readily emerge.

Sometimes dissension and bad feelings just linger on, especially in a house with strong personalities or people going through the more challenging times of life: toddlerhood, teenage years, mid-life burnout! Touching is remarkable in dissolving bad feelings—squeezing a hand, giving a hug, massaging a back. My son and I have very different personalities, and too often our conversations are less than congenial. But at the end of a conflict-laden day, I can usually call up the words of St. Paul that I heard often in my childhood: "Never let the sun go down upon your anger."

> *What is one gesture of balance or reconciliation you could do with a loved one? Do it now.*

Through babyhood and childhood I gave my son back rubs to put him to sleep. When he blasted into teenage years and abrasive words passed between us during the day, he would still ask for a back rub. Sometimes my response would be, "After the way you treated me, you want me to give you a back rub? Fat chance!" I would huff away, and later feel terrible. But when I massaged him anyway, despite our tiff, it softened his mood and mine and eased our hurt feelings.

Sometimes it's not easy to extend ourselves to another in touch—it can require our last reserve of energy. But the nonverbal is so soothing, balancing, and healing. Ashley Montagu's groundbreaking book *Touching: The Human Significance of the Skin* explains why. The skin is the largest sense organ, and the more stimulation it receives, the happier and healthier we are. Even the brush of wind against skin invigorates body and mind. Touch from another person defuses alienation and increases feelings of comfort and peace. Babies' internal organ systems are not fully developed at birth, and massaging them helps their digestive, respiratory, and excretory systems come to peak functioning.

Another form of nurturing stimulation is sharing food. Our lives are so busy that even sitting down and having meals together is a luxury that has to be planned and scheduled. Team practices, music lessons, evening classes, workout times, and professional meetings all pull us in separate directions. Still, a primal urge calls us to come together, to sit down, to shake off the frenzied energies, and to quietly talk over the day's stories and strains, victories and quests. Something on the table from the natural world—a fresh flower, a vase of autumn leaves, a candle—helps us clear our minds and calm ourselves.

What types of rituals call to you? What ritual have you wanted to start in your family? Initiate it this month.

When I think of it, I pull out notecards with quotes I've squirreled away over the years, and place them in cardholders in front of each plate. Over the course of the meal, we talk about what they mean to us. They evoke pleasant memories or congenial disagreements and take us on pleasurable tangents that keep the conversation flowing through dessert and dishes.

Something we have less control of, which seems to challenge us at every turn, is the troubling barrage of sex and violence to which children are subjected from television, videos, and movies. An incident from Suzanne's life highlights the dilemma.

Shortly after its release, Suzanne and her husband went to see the Clint Eastwood movie "Absolute Power." Knowing a little about the content of the film in advance, they were dismayed to see a couple bringing three young children into the theater. They were astonished when the two adults left the children in a row near the front and walked out. It was difficult for them to watch the previews, which contained shattering scenes of gun violence, explosions, and gore, and then the main attraction, without wondering about the effect they were having on these young children.

Throughout the movie, Suzanne found herself trying to understand the parents' behavior. Did they have any idea of the movie's content and the kinds of previews that would go with it? Were they using this movie as a babysitter so they could see a different film? Suzanne found herself getting more and more angry. Should she talk to the parents in the lobby afterwards, even though the children would probably overhear? Should she just forget about it? When the movie ended, the children quickly vanished into the crowded lobby and Suzanne lost the chance to talk with the parents. Although the appropriate action was certainly not clear, she felt an important opportunity had been missed.

When you see something that feels gut-wrong but that may not technically be your business, what do you do about it? Think of something that could make a positive impact and do it today. It could be writing a letter to a multiplex movie theatre about the movies or previews they show without monitoring the ages of the viewers. Or it might be talking to your neighbor in a nice way about some great parenting classes.

Unfortunately, children watching images inappropriate for their age and maturity level is all-too-common today. Perhaps the parents in Suzanne's story were badly in need of some time to themselves, and if they had had someone familiar to leave their children with would not have exposed them to such brutal images. Parents cannot possibly meet all their children's needs at all times, but when they can't there should be others around who can offer kindly and attentive care.

Tell a friend (or neighbor) who looks as if they need a break that you'd like to take their children on an outing. Or, you could offer to spend a fun evening at their house playing games with the kids so your friend can escape for a few hours.

In the twenty-one years since my son's birth, I have actively arranged suppers at our house and outings with other individuals and families with whom I wanted my children to be close. Some are good friends who now have children of their own; some are the families of my children's best friends; some are adults with no kids of their own who happen to love children. They touch my children's lives in unique and wonderful ways, exposing them to hobbies, interests, sports, or intellectual pursuits different from my husband's and mine. I've also gone out of my way to create friendships with older couples in our neighborhood. When my one last nerve has been stretched to the breaking point, my children have had many choices of people to call or hop on their bikes and visit!

Who would you like to draw into your sphere as "extended family"? Plan a simple supper and invite them over for the evening.

I couldn't wrap up these thoughts about family time without including a piece written by one of the wisest, most broad-minded, least judgmental people I know: Father Tom Seagrave. I tucked away these words from his 1992 Christmas card, marking the occasion of his twenty-fifth anniversary in the priesthood. I have kept it in my top desk drawer so I would come across it now and then. I hope you will cherish these jewels as much as I do; they are a touchstone for me in this high-stress world.

> Hitting people is never a good idea.
>
> Yelling at people is never a good idea.
>
> Violence in any form is never a good idea. Violent actions, words, thoughts, and feelings are much to be resisted from the heart.
>
> "Bad language" really is bad language; we use it precisely because it has "punch" ...and we use it to punch people. Bad language is violent language and violent language is never a good idea.
>
> Intimidation is never a good idea. It is the common methodology of the bully, the self-righteous and the ignorant, and the intimidator is almost always all three.
>
> Most "bad" things are done by perfectly good people. Most people do bad things *not* because they are lacking in personal goodness, but because they are sometimes foolish, sometimes

weak, sometimes desperate. Judgmental people almost always miss the point.

We all need to be easier on people. We have to cut a lot more slack for our families, our friends, and ourselves. "Lighten up!" is almost always good advice.

The Golden Rule is still the golden rule. I don't know anybody whose wish list includes being abandoned, betrayed, debased, discarded, replaced, nagged, overlooked, left-out, or treated as a non-person. If you don't want it served to you, then don't dish it out.

Questions For Reflection

- Think of a time when what you *said* to a child in your life was different from what you actually felt, thought, or did. What happened?
- Do you know a child who needs a little more time with an adult? Would you be willing to include her or him in your activities?
- What constitutes your "family"? Could you broaden your definition and in the process enrich your own life by including others who would like more involvement with you?

Chapter 13

Meaningful Work

Ellen Schwartz and Suzanne Stoddard

livelihoods both personally satisfying and earth-friendly

The wholesale substitution of machines for workers is going to force every nation to rethink the role of human beings in the social process.
—Jeremy Rifkin, *The End of Work*

Too many organizations ask us to engage in hollow work, to be enthusiastic about small-minded visions, to commit ourselves to selfish purposes, to engage our energy in competitive drives. Those who offer us this petty work hope we won't notice how lifeless it is.
—Margaret J. Wheatley and Myron Kellner-Rogers, *A Simpler Way*

It is not hard work which is dreary; it is superficial work.
—Edith Hamilton

Our human destiny is inextricably linked to the actions of all other living things. Respecting this principle is

the fundamental challenge in changing the nature of business.
—Paul Hawken, *The Ecology of Commerce*

In work, do what you enjoy.
—Lao-Tzu, *Tao Te Ching*

Work is love made visible.
—Kahlil Gibran, *The Prophet*

Today's young people see and absorb more about adult life than we think. They notice that grownups often have jobs they don't enjoy or feel proud of. Or they see that, while the work itself may be well-paid or interesting, their parents have to spend more hours at the office than ever because they are doing the extra work of someone who was downsized.

Then there's the larger group of people employed at the low end of the payscale. Many of them have to work two jobs just to keep the rent paid and the car running. Either way, children live the effects of discontent at work: little time with the people who mean the most to them.

Is it possible that we could have jobs offering more satisfaction than stress? Could we look forward to going to work and feel good about what we do for a living because it benefits others and doesn't toxify the planet? What would it be like to work fewer hours and have more time for ourselves and our loved ones?

When every day is scheduled to the max, we hardly have a moment to even imagine such a world. Yet new models of work are emerging, and more and more people are finding that with creativity, flexibility, and persistence they are able to follow their heart's desire and keep a roof over their heads at the same time.

Creating New Paths

People who have found (or created) work they really enjoy often have their feet in a couple of worlds.

We'd like to tell you about two individuals we know who have arranged their lives to have an adequate base income while spending the majority of their time doing what they love. One is Sammie Lee Hill of Richmond, California, and the other is Kimberly Weichel of Tiburon, California.

Sammie Lee Hill is a songwriter and actor, having changed careers in his early forties after being retired by the San Pablo Police Department because of injuries sustained in the line of duty. From the time he was eight years old, Sammie knew he wanted to be a policeman and he loved to write, especially poems and song lyrics. Although a mentor he had as an adolescent encouraged Sammie to go into law enforcement when he was eighteen, Sammie never stopped writing songs for fun in his free time.

After the injuries forced retirement, Sammie was depressed for a year, wondering what to do with his life. He enrolled in a paralegal certificate program, but when he completed it realized his heart was no longer in criminal justice work. A friend phoned him out of the blue and gave him a number to call in case Sammie wanted to apply for a job as a security director for a movie, "The Boys' Club," being shot in San Francisco. Not only was Sammie hired for the security director job, but he also got a small part in the movie as a policeman.

The movie's musical director, Dean Andre, heard that Sammie was a songwriter and said "Maybe we could collaborate sometime." Sammie said yes over lunch, but thought "Yeah, right—*him* call *me?"* Six months after the movie was finished, Sammie did get a call from Andre, who said he needed a love song as soon as possible and wondered if Sammie had some lyrics. Sammie said he didn't but that he'd work on it. Thirty minutes later Sammie called Andre back with words he had just written. Andre thought they were wonderful and arranged to meet Sammie to record the song. Sammie scheduled a studio and three back-up singers, and the recording session was done within two days. Now Sammie gets to write songs all the time.

We first met Sammie a couple of years later through one of the mentors in *Invest in Kids* (our mentoring and scholarship program). Sammie originally wanted to mentor a sixth-grader in the neighborhood where he grew up, but then realized that because his work took him all over the country he wouldn't be able to provide consistent attention to one child. He did, however, offer to write a theme song for *Invest in Kids.* We said great, and a few months later got a phone call that the CD was ready. When we listened to the finished recording of "Mentor," we were speechless. We cried and hugged each other, simultaneously saying "Yes!" Ten percent of the proceeds from this commercially released song now go to scholarships for our poverty-level, at-risk youngsters.

Kimberly Weichel, the other exemplar we offer, has created her own meaningful work while managing to sustain herself in one of the most beautiful places on earth. Nestled on the San Francisco Bay, Tiburon appears to be a place where only six-digit income earners live. But although Kim and her husband Carl both work, they are not wealthy. They rent out the back of their home, which they remodeled to be a full apartment, as a way to help pay the mortgage.

While Carl runs his direct sales business out of their home, Kim works three days a week at the Center for Spiritual Democracy as its part-time executive director. The Center is a nonprofit organization dedicated to reconnecting people and society with the spiritual values upon which America was founded in order to reinvigorate our democracy. Her job there lets Kim use her accumulated skills, contacts, and extensive background in the international arena, while challenging her in new ways to approach work spiritually. This salaried work complements the other pursuits to which Kim is deeply committed—her unpaid "work in the world." She heads the Marin chapter of the United Nations Association, is a career counselor for the Alumnae Resource Center, and serves as a lay liturgist at her church. She also writes about international issues for local publications and has published a workbook called *How to Nurture Spiritual Values in the Workplace.*

Kim and Carl relish their family time, taking frequent camping trips and embarking on overseas adventures at least once a year. Weekends usually find them on an invigorating hike or cross-country skiing jaunt with their son Julian. Kim says that once Julian was born she was highly motivated to find ways to spend less time at work and more at home. She and Carl both know that these are critical years in their son's life and are able to make time to help him with reading and schoolwork. Slowing down their work schedules to a less hectic pace has opened up the possibility of adopting a little girl, something they couldn't have imagined before. And it has enabled Kim to sing with a group she loves, "The Amazing Graces."

Both Sammie and Kim are in their mid-forties and they've acquired enough experiences to know what works for them and what doesn't. They believe in giving back to their community, and they find (or rather, create) the time to do just that. Their lives show that once people make a choice to do what they love rather than what seems safest, the universe opens up doors they never before imagined to exist. A decision made with both heart and mind seems to shift everything.

From Machine Era to Green Era

Matthew Fox, in *The Reinvention of Work,* describes the current paradigm shift from what he calls the Machine Era to the Green Era. In the mindset of the industrial revolution, life was depicted using the metaphor of a machine. All reality and experience were secularized; rationality and efficiency reigned supreme; the earth was a resource to be exploited; and decision making was hierarchical.

What is emerging now that is so very different is a consciousness of interrelationship and the necessity for tapping everyone's wellsprings of creativity. As we become more aware of the Earth as a living system, wonder and awe at the mystery begin to rise above doubt and cynicism. The power and beauty of learning is not in memorizing facts but in active participation; it is no longer the pursuit of mere knowledge that matters, but the seeking of wisdom. With all reality and experience sacred, a new hunger for myths abounds. A "job" results in a paycheck, but meaningful work brings out our special gifts and co-creates the unfolding universe. In partnership and in interconnection, we discover the art of living fully.[63]

Economic suffering and anxiety have always touched the poor, but with today's highly volatile global economy creating the possibility of stocks tumbling precipitously at almost any time, nearly everyone worries about the possibility of "losing it all tomorrow." As in all times of crisis, we have both the opportunity and the necessity for a radical re-visioning of our own lives, our culture, our economy, and of the earth herself.

E.F. Schumacher was one of the seminal thinkers in this shift with *Small is Beautiful: Economics as if People Mattered.* Peter Gillingham was privileged to spend several months with Schumacher during the last three years of his life, sharing his deepening and evolving thoughts. It was clear to Schumacher that "There is no outer prescription, course, or exercise that will bring a person to good work. We find it from the inside out.... Every one of us has the responsibility and the capability to construct his or her own map of reality and to dream, think, and act on the basis of it—then reflect on the action and its consequences to strengthen and make more fruitful the succeeding cycles of dreaming and thinking and acting." In his last year of life, he mentioned several times "If you know something, really know it, and don't act on it, then it will go sour on you, it will fester inside you."[64] Schumacher's huge vision is now being realized in a growing number of industrial concerns where serious attention is being given to the effects of manufacturing processes on the environment.

Spontaneous Synergy

Paul Hawken's book, *The Ecology of Commerce,* is replete with examples of businesses that have found ways to use clean technological processes to enhance the quality of life in the surrounding community. Although they don't get a lot of publicity in the mainstream media, many companies have already made their production processes less toxic. One case study involves the leading-edge company, 3M:

> In 1975, Joseph Ling, head of 3M's environmental department, developed a program called Pollution Prevention Pays (3P), the first integrat-

ed, intercompany approach to designing out pollution from manufacturing processes. The plan created incentives for the technical staff to modify product manufacturing methods so as to prevent hazardous and toxic waste, and to reduce costs. By reformulating products, changing processes, redesigning equipment, and recovering waste for re-use or recycling, 3M has been able to save $537 million. During a fifteen-year period, it reduced its air pollution by 120,000 tons, its wastewater by 1 billion gallons, and its solid waste by 410,000 tons. Over 3,000 separate initiatives contributed to the cause.[65]

3M is an inspiring example of one company; even more remarkable is what is happening in a whole town. In Kalundborg, Denmark, a coal-fired power plant, an oil refinery, a pharmaceutical company, a producer of sulfuric acid, the municipal heating authority, a fish farm, and some greenhouses and local farms are working cooperatively on waste re-use and recycling. The power plant started things off in the 1980s by recycling its waste heat in the form of steam. It had formerly condensed the steam to turn it back into water; it began sending the steam directly to the refinery and the pharmaceutical company. The power plant also found it could provide surplus heat to the greenhouse, the fish farm, and the residents of the local town, allowing 3,500 oil-burning heating systems to be shut off.

The refinery produced surplus gas, which was not used prior to 1991 because it contained excessive amounts of sulfur. By installing a process to remove the sulfur, a cleaner-burning gas was created, allowing a savings of 30,000 tons of coal at the utility company. The retrieved sulfur is sold to a chemical company. The process, which removes the sulfur from the smokestacks of the power plant, also yields calcium sulfate, which is useful to the sheetrock factory. Waste heat from the refinery warms the waters of a fish farm that produces 200 tons of turbot and trout, while the fish sludge is an excellent fertilizer for local farmers.[66]

This synergy is remarkable because it happened "spontaneously," without government regulation or law. Each exchange or trade was negotiated independently. Imagine what could happen in this country if an industrial community, or any community, were planned from scratch to maximize symbiotic relationships.

Some of the key principles Hawken suggests we keep in mind as we look to create new work are:

- Present-day limits need to be seen as opportunities.
- Degraded habitats and ecosystems must be restored to their fullest biological capacities.
- Having fun, being engaged, and striving for aesthetic outcomes create possibilities for dramatic, positive cultural shifts.[67]

Treading More Lightly on the Earth

Too much of our present production and consumption is for poorly made products that we don't really need. Yet the opportunities for useful production are many. Millions could be employed making solar technologies for new houses, where installation for heating and cooling is cost-effective, with conventional energy as a back-up system. Millions more could retrofit buildings with energy-efficient equipment. We could build rapid and convenient urban mass transit, including a network of bullet trains to speed across the country.

Amory Lovins, founder of the Rocky Mountain Institute, has proven beyond a doubt that we already have most of the energy-saving technologies in place to radically decrease our consumption of oil while maintaining a comfortable lifestyle. From factory production of goods to electric vehicles, all we lack is the political will to bring about a cleaner, safer world.

In a place called Gaviotas, Colombia, a man named Paolo Lugari imagined a community in which people would build a sustainable economy in a barren, rain-leached area where political and environmental crises had already wrought havoc. Initially begun by a group of technicians, indigenous Guahibo Indians, peasants, artisans, scientists, and formerly urban street kids soon found themselves engaged in exciting and satisfying work as they created cost-effective ways to meet basic necessities. Windmills sensitive enough to convert light tropical breezes into energy, solar collectors that work in the rain, and soil-free systems to raise medicinal plants and food crops now allow a community of several thousand to flourish. Does Gaviotas feed the spirit, too? Yes, with an acoustically exquisite concert hall within a sanctuary of verdant woods. Alan Weisman documents the astounding story in his book *Gaviotas: A Village to Reinvent the World.*

If this can be done in a Third World country, why can't even more remarkable innovations occur here? With no hesitation, we poured billions of dollars into the Manhattan Project to create nuclear weapons of mass destruction. Why not apply equal fervor and funds to renew America? We could create electric and solar vehicles so people could get themselves where they needed to go while having a far less destructive impact on the planet.

We could resurface roads with old tires, which would last longer than pothole-prone asphalt and be more soothing to drive on. We could manufacture pollution prevention equipment to be used at the point of production, in factories. We could clean America's rivers, one-quarter of which are unsafe to drink from. We could fund Lawrence Livermore National Laboratory to come up with technologies to break down our most toxic substances.

We could act as if the planet were our life-support system (which, of course, it *is!),* by paying people to rake leaves manually rather than using leaf blowers that add to the carbon dioxide buildup in the atmosphere. We could enjoy the feel of the earth and weed our gardens and lawns by hand instead of using carcinogenic pesticides that linger for decades.

Make your feelings known! Talk to others about using energy-saving technologies on a small or large scale. Write a Letter to the Editor. Contact your congressperson. Do what you can at your own workplace.

Preparing Today's Youth for Meaningful Work

Our educational system needs to be reinvigorated with visionary teachers and administrators, so that every child is well-supported by caring and wise adults. In ancient cultures, and even today in many countries, teachers are held in the highest regard. There, teaching is considered a profession requiring great talent and creativity, and it is well-paid. We could do far more as a society to encourage talented individuals to become teachers and prepare them better for the highly diverse classrooms of today.

Never more than now have we needed wisdom and compassion woven into the school curriculum. What if our schools, as a matter of course, fostered exploration, cooperation, imagination, and creativity, as opposed to the mere acquisition of knowledge? What if schools were really about tapping into children's vast potential and giving them tools for lifelong learning? We need to bring a sense of wonder back into our schools by making them havens for learning and truth-seeking.

We neglect our youth when we fail to teach (and to model) reverence for the natural and creative worlds and how to be part of a community. Children need to learn to work with others for results that benefit everyone. In this age of intense competition, fragmentation and alienation—aided and abetted by the frenzied corporate rush for more markets and profits—our children need a safe place where they can learn, grow, and thrive at their own pace.

Funding Jobs That Change People's Lives

Geoffrey Canada offers many suggestions in *Fist Stick Knife Gun* for work that brings meaning to people's lives while making an important contribution to the community. Canada proposed that we start a peace officers' corps that would pay welfare recipients a living wage with healthcare benefits to keep the peace and promote community pride. They would be trained in conflict resolution skills so they could go to the funerals of drive-by victims and defuse the revenge plots hatched in funeral parlors. As most of them are mothers and fathers, they would not hesitate to put themselves between the drug and gun dealers and their children.[68] They could organize youth to help seniors with yard work and house painting and elders to go into the schools and sit with their arms around the children, listening to them read. They could bring in university extension master gardeners to work with neighborhoods to create community gardens.

Jeremy Rifkin, in *The End of Work,* likewise suggests that offering a "social wage" in the nonprofit sector for performing essential community service work would be a viable alternative to welfare for millions of America's poor, benefiting both the welfare-to-workers and their communities. Rifkin calls this potential arena for meaningful paid work the "third sector" or the "independent sector" to differentiate it from government and business. In France, this social economy accounts for more than 6 percent of total employment, with 43,000 voluntary associations having been created in one recent year.[69]

Nearly two-thirds of inmates in California prisons have a learning disability. What if instead of paying $34,000 a year for each of them to be incarcerated, we would pay for aides to help children with reading comprehension in the early grades? A youth's identity could then come from his accomplishments and skills, rather than from the gold chain around his neck and the gun in his pocket.

The $125 billion spent each year on tax breaks and direct payment subsidies to transnational behemoths[70] would fund several million new service jobs. Another possible source is the biggest black hole for our tax dollars, the Defense Department. Lawrence Kolb, Assistant Secretary of Defense during President Reagan's first term, notes "Today the United States spends more than six times on defense as its closest rival, and almost as much on national security as the rest of the world combined." Many newly retired generals forcefully proclaim that we could cut the military budget by 25% and still have the best military in the world.

> *Call or write your congressperson today in support of a social wage to create desperately needed jobs that perform vitally important services. (See addresses in Resources, Chapter 6, at the end of the book.)*

Humanizing Healthcare and Removing "Litigious" from Legal Services

Our healthcare crisis is fertile ground for the confluence of creativity and compassion. Thirty-nine million Americans have no healthcare at all, and another 20 million have only minimal coverage. Even those who like their doctor find the costs to be exorbitant. The United States and South Africa are the only industrialized countries without guaranteed basic healthcare for all citizens. Every nurse and doctor we know considers profit and healthcare incompatible. With healthcare a corporate industry today, the individual patient too often loses out to the bottom line.

We must particularly examine the care of the dying. Almost two-thirds of our healthcare dollars are spent on the last nine months of life. Home visits by nurses and hospice staff are much more cost-efficient, and the peace that the dying person feels in the comfort of their own home, with family and friends for company day or night, is unquantifiable. Many new hospice staff have had divinity school training and are at ease talking to the patient and family about the passage they are approaching.

Almost every profession today needs to be infused with visions of broader potentialities. The combative structure of our legal system, for example, is enormously expensive and produces "solutions" that leave more rancor and revenge than satisfaction in the disputants. The increase in mediation and conflict resolution is a promising part of the new paradigm and needs to be made available everywhere disputes arise.

More Time, More Meaning

It is obvious that working people today need to withdraw from oppressive workloads and long hours. The grueling schedules we keep today have not been the norm for most of human existence. Only in the past 200 years has the arena of work taken such a big chunk of people's time. In other cultures and eras—even during the planting and harvest season—plenty of time was given over to weddings and religious rituals; in winter months, feasting and ceremonies could last two weeks at a time. It is unnatural for work to consume so much of our life force, leaving shreds of time for important life events. Until the Family Leave Act was enshrined in federal law in 1992, most working people were offered but a few days to take care of family crises or bond with new babies before coming back to work.

European workers, who are allowed four to six weeks of vacation a year, are pushing hard for the 30-hour work week with a living wage. Let's add our voice to the ground swell. How do we do this? By writing Letters to the Editor, by raising the topic in the office lunchroom, and by persistently calling and writing our legislators. With robotics taking over so many familiar jobs, the 30-hour work week is not a Utopian ideal, but an urgent necessity if we are to avoid epidemic unemployment.

We now have an incredibly stressed middle population working excessive hours while youth loiter on street corners and people over fifty can't find a job. CEO salaries and corporate profits have risen astronomically over the past decade. Surely these companies can afford to pay benefits for two people to share a job so both workers have time for a personal life. We must insist that corporations put family values on their agenda. After all, what does "family values" mean if not enabling people who love and care for each other to spend more time together?

Contact New Ways to Work in San Francisco at 415-995-9860 for many exciting work options you may not have thought of, including job-sharing.

Co-Creating the Support to Develop Work That Matters

We all need some combination of work that is both personally meaningful *and* beneficial to people and other living things. We should be able to have this without needing to worry about what would happen if we or a family member fell seriously ill and incurred a huge medical expense. Canada has a highly effective national healthcare system that frees people from having to take a job solely for the healthcare benefits and liberates retirees from agonizing over how an unexpected, debilitating illness could cause them to lose their home and be a burden to their children.

Call up your local, state, and national representatives and ask why they are not pushing for an equitable national system of healthcare where health coverage is not linked to a job. Set up a meeting with friends and co-workers at your legislator's office, and ask him or her to put your needs before those of insurance industry lobbyists.

Even the hourly contract work that is often the only kind of employment available today could be more palatable if healthcare were available from some other source than a 40-hour-per-week job.

While entrepreneurial skills are highly valued in the job market at this time, it isn't right that people who lack the ability to market themselves should be shut out of all but a few jobs. A full range of work options needs to be available to match the extraordinarily rich patchwork of talents people possess.

Some people find that their existing workplace isn't so bad, but they hunger for the opportunity to express more of who they are at work. One way employers are trying to satisfy this need is by encouraging employees to personalize their work spaces with cherished photographs, objects from the natural world, paintings, posters, or hobby items. Some companies and departments start meetings with a moment of silence or a brief meditation. Pacific Gas and Electric (PG&E) allows employees to receive their normal salary while spending four hours a month mentoring in the schools, as long as they match it with four hours of their own time.

Chris Chouteau, a department manager at PG&E, says "People have this illusion that managers have power, that we can create community. In fact, we can model what we want and try to lead, but we have little control over the outcome. We have to let go of needing to control the outcome.... We pretend the workplace involves only one level of a person ... but I need to be willing to take the risk to engage my work on more than one level.... Sometimes in meetings with higher-ups it seems to take a lot of courage to do this ... but the only thing that's really worthwhile is the engaging and connecting...."

With that in mind, Chouteau organized a retreat for his staff at the ocean where the employees were divided into teams and asked to comb the beach for objects they would use to create something. What they built collectively was to be a metaphor for whatever the employees wanted to bring forth at the office. When the whole group reassembled, Chouteau was profoundly moved by each group's creation, because it showed both the hunger for meaning that people felt and the deep creativity that wasn't being tapped in the current work environment.

Many of us are teeming with ideas about what we would ideally like to do with our time and our lives. For others, it's hard to imagine a work life outside of traditional job descriptions. Even if we have no idea right now what exactly that might be, we don't need to feel defeated at the start. To take the first step, we just need to identify one aspect we love, one thing that brings us joy, or one activity that excites us. In their book *True Work,* Michael and Justine Toms give several recommendations for finding your own life's calling: • Take the initiative

- Keep up your energy and enthusiasm
- Set an intention
- Hold the future loosely—don't be rigidly attached to a specific outcome
- Persevere!

The Tomses consider that mistakes and disasters are actually "directed crises," and that there is wisdom to be gleaned from every part of the life experience.

The Solutions Are Ours to Create

We do not live in a world of quick and easy answers. There is no all-knowing guru to tell us what the rules are. There is no pat formula for success and happiness. Yet we are privileged to live in times when average people can initiate actions that give them more control of their destiny while helping to move their communities forward. What is deep in our hearts connects with the unfolding of an ever more rich and magnificent universe.

Write down ten things you've done in your life that have given you a deep feeling of fulfillment. Draw from small and simple, as well as major, tasks and accomplishments. Include as much detail as you can recall. You'll be amazed to see how one or more points to something you could be earning a living at or doing regularly for great satisfaction and a partial income.

Countless people who want to turn their creative passions into jobs—to be artists or writers, business coaches or teachers in innovative schools—are unable to find a way in our current economy to do this and still support their families. Many have buried their dreams, extinguishing the most vital part of themselves, and settled for long commutes, co-workers whose values they don't share, and being half the person they could be. In the most extreme cases, this denial of one's inner calling leads to alcoholism, severe depression, abuse of family members, and even suicide. As a society, we cannot afford to let this monumental amount of creative passion and talent go to waste.

When we attune more to others, asking questions that encourage deeper delving into dreams, needs, and desires, we support each others' wholesome hunger for a bigger life. By the simple act of asking "What do you really want?" "How do you really feel? or "How can I help?" we take the first step toward affirming their inner yearnings, enlarging their window on the world, and awakening to our own emerging possibilities.

Questions For Reflection

- In work situations, what stresses you? Where do you feel compromised? What gives you satisfaction? When do you lose track of time? What brings out the best in you?
- What would your dream job, career, or livelihood look like? How many hours per week would you work? How would you be interacting with others in alignment to a purpose? What would you really like to do with your time and your life?
- How does your urge to create beauty surface?
- What "limits" could you turn into opportunities? What key to freedom is hidden in a structure of control you're currently oppressed by?
- In what ways can you bring more fun and engagement into your life?
- What ideas do you have for radically re-visioning your profession?

Chapter 14

Giving Time, Getting Joy

Suzanne Stoddard

life as a banquet for the servers

I slept and dreamt that life was joy. I awoke and saw that life was service. I acted and beheld that service was joy.
—Tagore

Only those among you will be truly happy who have sought and found how to serve.
—Albert Schweitzer

If you look closely you will see that almost anything that really matters to us, anything that embodies our deepest commitments to the way human life should be lived and cared for, depends on some form—often many forms—of volunteerism.
—Margaret Mead

Every day I am reminded how much my own outer and inner life depends upon the labors of my fellow men. I must exert myself in order to give, in return, as much as I have received.
—Albert Einstein

You give but little when you give of your possessions.
It is when you give of yourself that you truly give.
—Kahlil Gibran

Ever since I was a child, I've found enormous satisfaction in helping other people. I began to do volunteer work because I felt better about myself when I was doing it. I became a regular at volunteering, not because I am a selfless ascetic but because I get so much out of it! I am addicted to enjoying life and learning new things, and volunteering has never let me down in these areas. I have developed skills and talents through volunteering, felt needed and useful, and been privileged to meet extraordinary individuals. Several jobs have emerged for me as a direct result of doing volunteer work. Many of my most solid friendships have been formed with people I met in the course of community service. The quality of people who have entered my life as a result of volunteering in peace issues, particularly, has been sublime.

Shattering Stereotypes

In 1998 I joined a support group for inmates at San Quentin Prison after attending a nonviolence workshop sponsored by the Alternatives to Violence Project (AVP). The last Friday of each month now finds me with a group brought together by their common desire to internalize peaceful means of resolving conflicts and to radiate peace outward. The San Quentin contingent includes many "lifers" who have joined AVP and attended its trainings in order to come to terms with the violence they have both perpetrated and been victimized by. Some of the non-inmate volunteers have also had direct experiences with violence. Those of us from the "outside" are there to support the inmates in their process of creating calmer and more peaceful selves and relationships. After all, what community needs nonviolence training more than those inside prison walls?

With my very first time on prison grounds, stereotypes I had long held about "criminals" were shattered. I was amazed by the articulate, thoughtful words of many of the inmates. Weren't 85% of prisoners functionally illiterate? How could they speak so eloquently, so obviously from the heart? In this Quaker process, everyone does a check-in at the beginning and talks about anything that's been going on in their lives that they want to share with the group or ask for support in. There is no side-talking allowed, thus whoever has the floor is given total attention. One by one, the men revealed their humanity in proud or poignant stories of family visits, children going off to college, illnesses of loved ones, or sons and daughters who never call or acknowledge their incarcerated dad.

The most moving part of my initial visit to San Quentin was when an inmate named P.J., who had just had a birthday, stood in the middle of the circle receiving compliments. AVPers from the outside who had led workshops in San Quentin with P.J. praised his tireless efforts to promote the nonviolence project within San Quentin. But what brought tears to my eyes (and I had only just met P.J.) was hearing other inmates describe him as a "caring friend", a "positive light," and a person "who always had time for others in trouble." One fellow prisoner of a different race just looked straight at P.J. and said, "You're awesome, man."

The Power of Doing What We Don't Have to Do: Just Being Kind

How can something that expands our emotional life and our connections with others have an image of being onerous? Why do so many feel that volunteer work is a sacrifice or a painful duty, when it is far more compelling than nearly anything on television? Why spend so much time experiencing life second-hand, when you can have your own rich experiences? Community service is something anyone can do. We all have far more choices than those in prison but often live our lives as if we were in jail—limiting our activities to working, sleeping, eating, paying bills, doing what we have to do, and watching TV.

Helping those we are not obligated to help is a guaranteed way to generate fulfillment, new friends, greater self-confidence, and affirmation that one is spending time and life energy on something of value. How many other activities can we say this about? All too many people work at uninspiring jobs out of financial necessity, so it is incredibly satisfying to do something in our free time that we choose consciously, deliberately, and with no thought of material gain. It is a very positive, powerful feeling! And the need is huge.

We've all heard stories of ordinary people who became heroes during disasters, showing little or no concern for their own safety. While courage in emergencies is inspirational, that same caring and connectedness could vivify us every day.

We have been unwittingly seduced by advertising messages extolling comfort and personal convenience as what make life worth living. In modern American life, the value of kindness is not given its due. I love the bumpersticker *Practice Random Acts of Kindness and Senseless Acts of Beauty.* In volunteering, there is something for everyone—from singing at a convalescent hospital to planting flowers, from playing with tots in a crisis nursery to helping roof a *Habitat for Humanity* house.

Mentoring the Young—You Are Needed Now!

One of the most urgent needs for volunteers today is in the arena of mentoring youth. Mentoring is a rewarding way to help a young person who is facing challenging family problems including economic disadvantage, get on the path toward a responsible, purposeful life. This starts with helping instill good study habits. Mentoring involves no special skills and most organizations seeking volunteers provide training and support. All you have to do is be yourself. We often have no idea the gift this is to those who've experienced little discipline, structure, guidance, love, or consistency at home. It's shocking how many of today's young people fall into this category.

The heart of mentoring is believing in a child. A mentor spends a couple of hours a week with a young person, helping with homework, noticing budding interests and talents, and fostering problem-solving skills. When something distressing happens, youth often view the event as an unmitigated tragedy. When the mentor says "That was tough, but what else could you have done?" the question is usually met with a blank stare. As the mentor helps brainstorm possibilities, the young person sees a whole range of actions that could have made things turn out better. When the mentor then asks "What can you do now?", the child begins to understand that no matter how bad a situation is there are still many choices for getting things back on a positive track.

A mentor avoids telling a young person what to do, but by sharing experiences honestly and openly from the mentor's own life helps the youngster come up with an idea that relates to his own dilemma. Young people are especially surprised and encouraged by their mentor's stories where a "catastrophe" turned out to be the best thing that ever happened. Not having children of my own, I find that mentoring helps me feel connected to young people in a wonderful way. I can expose them to things their parents haven't the means or the time for and in the process I get to share things I love: book browsing, rollerblading, tennis, horseback riding, and baking cookies. Children today need more *real* experiences like these to counteract the time they spend passively in front of TV or playing computer games. In the process, they develop useful skills and healthy passions.

Know any single moms or dads who would relish sharing their child with another adult occasionally? Seize the opportunity to be a friend and positive role model to a youngster and discover how much fun hanging around with kids can be!

The Printed Word: Our Kids Need More Reading Help Than Ever!

Another substantial and enjoyable way to be of service to children is reading to them when they're small and letting them read to us when they're just learning how. Done on a regular basis, reading makes all the difference in a child's success in school and in later life. When children fail to become good readers, their progress in all academic subjects is severely limited and this lack of early success casts a long shadow on their future prospects for holding a job and staying away from drugs, crime, and dependency. Like almost no other activity, reading is a window on multiple new worlds, enlarging children's sense of who they are and what they can be.

With most parents having to work during school hours, teachers need other adults in the classroom to help cope with large classes of highly diverse ability levels and language backgrounds. Consider helping your local school with tutoring. If working directly with a child doesn't quite hit the mark for you, consider fundraising. Most of our inner-city schools are in desperate need of the money that is easily raised in affluent communities. Another community need is helping immigrant and low-income families fill out paperwork to enroll their children in Head Start—a program proven to establish a positive attitude toward learning in very young children. Excellent government programs like this often go begging for clients, not because the need isn't huge but because the intended recipients of the services aren't aware that they exist or don't have the education or proficiency in English to fill out the forms.

Can Imperfect People Help Others? Yes!

Nonprofits, daycare centers, and volunteer agencies treasure retired people for their stability, nurturing skills, well-honed talents, and time availability. The poet Robert Bly, who is himself a senior, says that our culture sends elders the wrong message. Instead of encouraging them to relax, watch TV, and play golf, we should be saying: Your kids are grown, you've made plenty of mistakes with them. Now go out and do a better job with someone else's children!

Many people today feel they can't go out and make the world a better place until they've "gotten themselves together." They spend substantial sums of money on therapy and massages or go overboard in analyzing every conversation that went awry. Don't get us wrong. We too love massages and have been avid journal writers. We've also gotten through some very painful times with the help of insights gained in therapy. But James Hillman, in *We've had 100Years of Psychotherapy—and the World's Getting Worse,* reminds us that none of us will ever be perfect and that targeted efforts on real-world issues can eliminate the root causes of many psychological problems.

Helping someone else is the best cure for the blues. Do something nice for someone today, even if you don't feel so great yourself.

Even on our deathbed there will be inner demons to duel and shadowy places untouched by the light of consciousness. We can't postpone the important work we need to do in the outer world until our personal weaknesses are overcome, because it will never happen. More important, we grow in strength, courage, and skills through tangible external accomplishments. George Bernard Shaw said it best: "This is the true joy in life, the being used for a purpose recognized by yourself as a mighty one. The being thoroughly worn out before you are thrown on the scrap heap of life, the being a force of nature instead of a feverish, selfish little clod of ailments and grievances complaining that the world will not devote itself to making you happy."

Deepening Your Travel with Service

If you love travel, nature, and new experiences, you will almost literally drool while reviewing Earthwatch Institute's page after page of exotic service opportunities abroad. Trips are often led by scientists and university professors doing field work in archaeology, biology, anthropology, sociology, or entomology. Here's a sample of trip names from the 1999 catalog: "Bees and Orchids of Brazil," "Icelandic Glaciers," "Folklore of Rural Russia, "Saving Philippine Reefs," "Sri Lanka's Temple Monkeys," "Old World Songbirds," and "Irish Music at the Crossroads."

You are likely to have as co-travelers a fascinating group of people from all over the world who, like you, are looking for adventure, excitement, and an opportunity to help on a research or service project. It is not necessary that you have any expertise. You will be given tasks to match your skills or interests once you arrive at the destination. (See Resources, Chapter 14, at the end of this book for contact information on

Earthwatch and other organizations offering service travel.)

Broadening Our Definition of Education to Include Service

A growing number of schools are allowing students the opportunity to volunteer in the community and learn real-world skills while helping others. Cornerstone in Vallejo, California trains junior high school students to help mentally and physically disabled children ride horses. In the process, the young volunteers learn how to saddle, bridle, feed, groom the horses and muck out their stalls. Youngsters build friendships with disabled kids, while learning useful and satisfying skills. They are rewarded for their work by being allowed to ride the horses during slow periods and gaining credits at school.

In the San Francisco Bay Area, hundreds of teachers are turning the outdoors into classrooms to teach science, math, art, writing, reading, and community service while helping wildlife thrive. Because of the efforts of students at a Petaluma High School, fish have returned to Sonoma County's Adobe Creek. Oakland and Richmond elementary pupils teamed up with the East Bay Regional Park District to clean up Wildcat Creek in Contra Costa County and plant steelhead trout they had hatched and raised in their classrooms.

If you are a teacher, parent, or simply an interested community member, promote the idea at your local school of allowing students school credit for doing a volunteer project. Real-world service learning is win-win!

Living Outside Ourselves, Modeling for Others—Starting Now!

Our own children learn an invaluable lesson when they see us helping others. Modeling the behavior we want from kids is always more effective than telling them what we expect them to do. In a hectic age when many of us wonder how we can make the world better, volunteering gives us a chance to temporarily step out of the rat race into a quiet place where time expands and we can enrich ourselves by touching the lives of others.

Volunteering can be especially valuable when our personal frustrations loom large. We may find ourselves feeling new gratitude for our own lives because we see that others, with more daunting problems than our own, find the courage to go on. And the appreciation we encounter can't help but boost our own spirits!

Calling the local volunteer bureau (generally found in the business section of the phone book under Volunteer Center of ______ County) will provide a kaleidoscope of helping opportunities. Want to engage in something outside the typical parameters of volunteering? If you're incensed about Nike paying 29¢ per hour to workers making athletic shoes that sell in the United States for $129, Global Exchange (415-255-7296) would love to channel your righteous anger into constructive action. Does it seem likely that the pesticides doused on our vegetables not only kill bugs in the farm fields but also leave long-term imprints on our immune systems? Call the Pesticide Action Network (415-981-1606). Are you concerned about chemicals acting in combination and increasing in toxicity in our air and water? Call the Toxic Links Coalition to find out what they're doing about it: 415-243-8373. Does your passionate spirit recoil at current assaults on hard-fought environmental laws? Call Friends of the River or the Sierra Club and get to work!

Find out about chemical pollution in your own community. By entering a zip code in The Environmental Defense Fund's website, www.scorecard.org,*you can view a map of local polluting facilities, find out more about the chemicals these sites emit, and get contact numbers for manufacturers in order to voice questions or concerns.*

You can teach someone to knit, paint posters for a fundraiser, or expose disadvantaged kids to things you love, whether camping, chess, carpentry, guitar, or shooting hoops. Homebound folks often just need someone with a car to run to the pharmacy or grocery store. Advocacy groups for battered women, orphaned animals, campaign finance reform, affordable daycare, or teen pregnancy prevention would be thrilled to have a few hours of your time! Recently I spent a total of four hours, on two occasions, tutoring an elderly, disabled immigrant so he could pass his test for U.S. citizenship, and unexpectedly made a sparkling new friend. The delight I have already received from his friendship far exceeds the effort I expended to help him.

Usually the hardest thing about volunteering is taking the first step. But if your time is really restricted, you don't have to commit to a weekly or monthly stint. Even one-time volunteers at the local public radio station, library, or cerebral palsy fun run are gratefully welcomed. The list of organizations willing to accommodate helpers of all commitment levels goes on and on. No doubt, they hope that once you try it you'll like it enough to come back, but there's never an obligation to do so.

The possibilities are endless. Once you've figured out what gets your juices flowing, make that initial phone call. You'll only regret that you didn't do it sooner.

Questions For Reflection

- Think of a time when you did something that was easy and fun for you, and also made a big difference in someone else's life. What is keeping you from doing this again?
- What interests, passions, or skills do you have that someone else could benefit from? (If you can't think of at least five, you are not trying hard enough! You can read, write, have a phone conversation, cook, be a friend, play a sport, drive a car....)
- Think about your life from childhood on. How has it been enhanced by volunteers in the schools, on sports teams, at church, or just by kindly folks in your neighborhood who've lent a hand?
- What is the one thing you like to do most? What ignites you with a sense of purpose and meaning? Could you do this for or with someone else?

Chapter 15

Together We Are Whole

Suzanne Stoddard

new ways to create a support network while beating the high cost of living

We are faced with having to learn again about interdependency and the need for rootedness after several centuries of having systematically—and proudly—dismantled our roots, ties, and traditions.
—Paul L. Wachtel

We cannot deny our connectedness as we build our separateness.... When we link up with others, we open ourselves to yet another paradox. While surrendering some of our freedom, we open ourselves to even more creative forms of expression. This stage of being has been described as communion, because we are preserved as our selves but are shorn of our separateness or aloneness.
—Margaret Wheatley and Myron Kellner-Rogers, *A Simpler Way*

Always expand. Never contract.
—Tarthang Tulku

I recently read a story about a woman who learned she was suffering from terminal cancer. After the shock, rage, and grief had swept over her, she called her teenage daughter and ten friends and co-workers to a bedside meeting. She told them she had only a short time to live and asked if they would be willing to be her extended family during the next several weeks. Her mother and only sister lived far away and were not able to be closely involved in her care. The woman said she would need a great deal of help, and asked if each person could find a way to play a vital part: giving her medications, handling her personal business, feeding and dressing her if necessary, and helping to make arrangements for her daughter's future. She gave everyone a chance to either accept or say no.

Some of the people had little spare time, while others wondered if they had the necessary caregiving skills. In talking it out, each thought of things she could do to help the woman. They did not necessarily feel adequate to the task, but each agreed to be a part of the woman's support system as she confronted her death.

Several weeks later, the woman passed away.

Rethinking Our Separateness

When the helping group reunited to talk about the commitment they had made, there was an outpouring of emotion. They spoke of the great honor bestowed on them in being asked to help ease their friend's last days. They all said they had found inner strengths and reserves of caring far beyond what they had ever called forth before. They had acted as a community with a clear and urgent mission and gracefully accomplished the purpose for which they'd been brought together.

When I read this, I asked myself, How often today are we directly offered the opportunity to give of ourselves so selflessly?

The tragedy is that few people expect this of us and generally we don't expect it of ourselves—or ask it of others. We live in a world where millions of people have little or no "community" and even those with friends and family nearby don't feel comfortable asking them to do something inconvenient, especially if it involves exposing dysfunctions, inner demons, or other "weaknesses." By not asking for help, we deprive ourselves of what we most need, to feel loved unconditionally by others when we feel least lovable. We deprive our friends and family of the chance to enlarge their own lives by being a more integral part of our own. Maybe they're just waiting to hear "I'm in bad shape and I need you."

If only my brother Rupert had known how to say these words, things might have turned out differently.

Troubled and Nowhere to Turn

On May 13, 1991, at the age of 40, Rupe entered his garage from the kitchen a little before 3PM, closed and locked the door behind him, attached one end of a hose to his truck's tailpipe, pulled the other end through the driver side window, rolled the window up as far as it would go, turned on the ignition, and inhaled exhaust fumes. By the time his 10-year-old daughter got home from school, he was dead.

The reasons my brother ended his life will never be perfectly clear to me. When someone does something so final, so desperate, there is rarely only one reason. If there is blame to attribute, I think it is to a set of cultural norms Rupe had absorbed that prevented him from getting the help he needed. I don't believe Rupe loved and valued himself as others loved and valued him. And I don't think he knew how to ask for help.

Rupe's notions of manhood were learned in the 1950s and 1960s when the extent of advice given by many American fathers to sons in distress was "be a man," "take it like a man," or "men don't cry." Asking others for help, confiding vulnerabilities, or talking things through were not the models our fathers had been taught, so they reflexively passed on to their sons this rigid, hollow model of how males should "tough things out." Unfortunately, telling others to be strong without offering specific strategies and tools often leads to denial of painful emotions that can fester and contribute to mental illness.

Like so many children of the 1950s and 1960s whose fathers had returned from the carnage of World War II with the driving desire to make everything safe for their families, we lived in a sheltering cocoon that was often happy but where bad moods and critical outlooks were judged rather than probed. In Rupe's early adolescence, when his behavior didn't reflect what Dad considered positive, Dad would say to Rupe "Why are you so negative?" A better understanding of psychology might have helped our father to be more constructive. Paraphrasing Thomas Carlyle: Tell a man he is negative, and you help him to become so.

The male bonding that is supposed to occur through watching endless hours of football and basketball together failed to forge meaningful father-son connections. When spectator sports fill most of the time spent together, how can a relationship not remain on a superficial level? What life skills are passed on? What deeper interpersonal understandings are developed?

It is a characteristic of depressed people to believe their own sorrows are a unique hell, and that what works for others has no application to them. While Rupe was very resourceful with his hands, his anger and his lack of skills to work through his problems cost him one job after another. His inability to provide adequately for his family apparently made him feel that his wife and daughters would be better off without him. Because Rupe had somehow learned that it was not okay to express personal weakness to others, he didn't know how to ask for help.

Had Rupe known how to ask and had our family not been so careful to respect his privacy, something would have broken open to make healing possible. Rupe would have felt part of a community of caring, and our support and love could have held him together until he was able to be strong again. We would have reflected back to him his generous nature and affirmed the joy he could put on his daughters' faces. But his depression (I see now) made him feel alien and exiled from his family and friends. He needed to take his medication *and* be able to accept help, and he couldn't, and he died as a result.

In death there are many lessons, and one is the reminder that there are suffering people who can be helped, and that we are called to help them. The death of a loved one is also a wake-up call that family (however we define the word) is infinitely precious and irreplaceable, and that invading someone's privacy can be the most loving thing we do.

Primal Connections: Family Matters

Family life is a blend of dark and light; mine has been no exception. There has been much joy, love, and laughter over the years, along with rituals linking us to the past and to each other. Some of my fondest memories of growing up are the hot dog barbecues with cousins, aunts, and uncles on Saturday evenings at my grandparents' house in the cooling twilight of a hot summer day. The best part was listening to the stories the adults told around the fire, and the catchy, corny songs they sang from their own early years, like "Don't Ya 'Buse My Darling Little Poodle" and

"Ivan Skavinsky Skavar." I couldn't have verbalized it at the time, but I was fascinated, as were my brothers and cousins, by this glimpse into our parents' past.

People of all cultures and times have told stories around campfires as a way to pass on their history and inculcate enduring values in the young. The stories my parents and relatives told held us rapt in the moment. I wanted those evenings to go on forever. My brothers and I would valiantly struggle against sleep so the magic wouldn't end too soon.

Today's families all too often rely on TV and videos as their primary entertainment source and they miss those satisfying interactions people used to have before technology allowed us to be so passive in our entertainment. Old-fashioned picnics in the park are rare for all but today's immigrant families, who seem to have a great time doing what most of us have forgotten. Water World and Disneyland can be fun once in a while, but they are too pricey for most families and don't exactly bring the generations together. (I have seen some of the most stressed and angry parents and whining kids at theme parks that I have encountered anywhere!)

What activity have you not done for a very long time that brings you peace, joy, or exhilaration? Could this experience include another person? A child? Do it this week.

Back to the Future

Why not combine the best of earlier times with the possibilities modern life affords? The essence of community is interdependence, and it can take many forms. Ellen once had an arrangement with a friend where on alternate Thursdays one would bring the other's family a complete dinner. Many families with young children find that creating a babysitting co-op offers respite to the adults and broader play opportunities for the kids.

> *Start a salon with eight or ten friends. Meet every month for a delectable potluck dinner and take turns choosing an article or a chapter of a book to read and discuss. In the summer months, go camping together or to a storytelling festival.*

While many of us prefer the intimacy of face-to-face contacts, millions of people are finding their lives expanded through virtual connections with strangers across miles of space-time. Via e-mail and the Internet, doctors, scientists, and college professors can share research and discoveries with colleagues in India, Argentina, or Finland at the flick of a button. American teachers share lesson plans and classroom management techniques with educators around the globe. Writers, lawyers, and students save time and gasoline by researching books, briefs, and term papers on the Internet. Infertile couples find a support system, as well as resources on adoption and in vitro fertilization on websites and in chat rooms. For people who are homebound because of illness or disability, the Internet can be a lifeline—not quite the same as a heartfelt hug, but a miraculous way to exchange words of comfort and diversion.

Community in the Workplace

Many companies are discovering that building community at work leads to a more robust bottom line. They are finding, as well, that the workplace becomes more creative and stimulating when workers do the majority of their work as part of a team. Group synergy leads to better products and more efficacious solutions. Employees are happier and more accountable, finding a higher motivation level in partnership than in solo effort. They feel both the responsibility to make an excellent contribution *and* the support of others who can complement their efforts.

In community-supporting companies, employees are expected to learn how to perform each other's tasks through training and apprenticing opportunities. People are allowed to grow and thrive when the organization defines roles but doesn't assign them to specific individuals. Some companies even rotate team leadership so that every member has a chance to learn and demonstrate newly acquired knowledge and skills.

Offering on-site childcare is another way for companies that value community to reap dividends in higher productivity and lower absenteeism. Moms and dads worry less and are relieved of a lot of schedule-juggling; and they can visit their little ones at lunchtime. Employers benefit from their employees' enhanced ability to focus fully on the task at hand.

If you work for a large company and there is no on-site child care, find out what could be done to establish it. This may seem daunting, but few worthwhile things get done without group effort. If you put together a committee of others who are interested, you can divide the various tasks according to peoples' interests and skills. You might be surprised that your boss gives you some release time to work on it.

When hierarchical strictures are loosened up, communication can flow in many more directions, to the benefit of all. Communication tends to be far more open in organizations that support community in their workplace. Workers come to feel comfortable expressing more of who they are and form deeper relationships with their co-workers. And because each team member's contribution is critical to the whole, each strives for high performance, which is then reflected in the end product. Comparing human systems to nature's ecosystems, Margaret Wheatley and Myron Kellner-Rogers have written:

> When individuals fail to experiment or when the system refuses their offers of new ideas, then the system becomes moribund. Without constant, interior change, it sinks into the death grip of equilibrium. It no longer participates in coevolution. The system becomes vulnerable; its destruction is self-imposed....[71]

They write that, in contrast:

> In ecosystems, members seem to have access to the whole system. The quality of their communication is dazzling. Birds build their nests over a river at different heights each year in anticipation of the coming flood levels.... How do they know this? We don't know. But clearly they communicate superbly. Nothing we humans have created in any human organization comes close.[72]

Paying more attention to nature, respecting her ways, and studying her deeply can only help us build workplaces where community and inclusivity thrive.

A New (Old) Way to Live: Cohousing

Though communes in the 1960s had a less-than-positive image in the mainstream media, a new development called cohousing or cooperative living is sparking interest in all age groups as a way to live more economically and sociably. In this type of arrangement, families or individuals own their own townhouse or apartment, which is part of a larger complex with a central greens for sunning and sports and common areas for arts and crafts, dancing, games, storytelling, and music making, and a fire-pit for twilight gatherings.

In cooperative living, children don't have to be driven across town for play dates. There's always someone to talk to, learn a skill from, or play ball with. A ready-made support system is there when you or your child are ill. Children can safely ride their bicycles on the path around the enclosed greens area until dark, while the parents enjoy a relaxing moment after work, free of worry. For many, a highlight of cohousing is the opportunity to share meals at the commonhouse kitchen, where rotating responsibility for cooking and cleanup means there are a lot of nights where you just show up and enjoy the meal.

Do you worry about being alone in your later years? Start talking with others to brainstorm a variety of options for a secure and affordable living arrangement. Taking consistent steps can get you there!

A highlight of mixed communities is the intergenerational mingling. Children without grandparents nearby like to hang out with the older folks. Retirees often enjoy helping with homework—they usually have more patience and time than busy parents! In cohousing, even teenagers accustomed to complaining of boredom find they have much to share with younger children, and they like being looked up to. Whereas play in same-age groups often has a competitive edge, children of mixed ages tend to play together cooperatively. In such a rich environment, television becomes the activity of last resort.

Try conversational evenings with one or more friends. Start with a key question, and enjoy a deepening connection. Ask: What was a turning point in your life? What do you see as the long-lasting influence of your father or mother? Who have been the other significant people in your life and what made them so? What would be a perfect day for you?

Not to be overlooked is the fact that cohousing allows people to work at lower-paying but more satisfying jobs due to economies of scale. Upkeep of common areas is handled by residents on a rotating basis, who consider it a small price to pay to keep the cost of living down. The time-honored barter system generally thrives in cohousing facilities (a massage for a haircut, a flute lesson for an oil change).

There was a time when people helped their neighbors raise a barn and kindly adults took a wholesome interest in other peoples' children. There is no reason we can't re-create a modern day version of that neighborliness. Safety is a key concern for most people these days, and life in cohousing communities *is* safer because residents take responsibility for each others' welfare. Neighbors don't stay strangers for long.

Is cohousing out of the question for now? What about staying planted where you are and following the lead of one urban block where neighbors turned their fences into firewood. They opened up all their backyards and a wide space emerged—for meandering trails with lots of benches and flower patches, and a large green field for games, sports, and cartwheels on the grass.

> *How could you stay where you are and make your neighborhood more connected and caring? Call in a neighbor or two and ask what they think.*

Initiating a more caring community is easier than you might think. You can start by bringing a fresh-baked pie or greeting card to an elderly neighbor, offering to watch a newly divorced friend's children so he can have a night off, or inviting a few coworkers over for coffee. Take the first step boldly, and watch the magic happen!

Does the idea of cohousing intrigue you? Call The Cohousing Network in Berkeley (510-486-2656) to find out what may already be happening in your area.

Get familiar with an enticing new website, www.favors.org,*that:*

- *identifies generous people offering services in the community, inspiring our youth with positive role models.*
- *identifies and supports nonprofits working to enhance community life.*
- *combines generosity and accountability by accounting on the web for favors people do for each other.*
- *provides easy access to a pool of skills, interests, and services available for barter.*

Questions For Reflection

- Does the idea of having more community appeal to you? In what ways could your time and energy be freed up?
- Do you have talents or interests that could be expressed more fully in fellowship with others? How might you start realizing some of those possibilities?

- Do you often wish there were more adults and playmates for your children to be around? Would this give you a respite, as well?
- If you don't have children of your own, would you enjoy being around kids a little more often?
- How might a serious problem at work have been ameliorated if you and your co-workers had met away from the office to discuss it and brainstorm solutions?

Chapter 16

Paring Down Our Lives

Ellen Schwartz

how less can be much more

Our life is frittered away by detail ... simplify, simplify.
—H.D. Thoreau

Simplicity isn't a sacrifice ... it's a gift. Simplicity isn't a moral achievement or a civic virtue ... it's a gateway into a free country....
—Mark Burch

Things that matter most must never be at the mercy of things that matter least.
—Goethe

Our lives are so full of the images of the good life: the big house, the sleek new car, the fabulous annual vacation and frequent weekends away, the feel of new clothes fresh off the rack, the taste of gourmet dishes at trendy new restaurants. What's not to like about this? Unfortunately, the price tag for such a life is a multi-digit income, generally from a corporate job involving long hours.

Some of my friends say "Ellen, how can you be so down on corporations? People need these jobs!" But what is it we really need? Do we need a bigger house? Do we need the pressure that makes our breathing shallow and our chest tight as we preview the day's agenda at 8AM? Do we need a new car when the old car still gets us where we want to go with far lower repair costs than a monthly car payment? Is where we are right now so bad that we can only find respite by running up a vacation on Visa rather than delighting in the dewdrops hovering on a leaf's end, or savoring the smell of a new recipe as it comes together in our kitchen? Do we need new clothes when the ones in our closet have neither jagged holes nor split zippers, and we're still gaining and losing the same five or ten pounds we've been struggling with the past fifteen years?

Being human, the new, the elegant, and the luxurious tantalize us. But the cost is high. The cost of being so wound up that we can hardly think of five things we're grateful for. The cost of being so rushed that even on our day off we don't have time to lie on our backs in the grass and watch the clouds roll majestically onward. The cost of moving so fast that we haven't a moment to stop and say a kindly word, or pen a few heartfully chosen thoughts on a notecard for someone who's hurting. The cost of not having the time to lift our faces upward and feel the soft rain. The cost of being too exhausted at the end of the day to read aloud to our child, not just storybooks in preschool days, but in middle school years, too, where the more nuanced books spark surprising comments offering windows into their souls.

Why are we so driven? Is what we are getting commensurate with what we are giving up? What is it that really gives us deep satisfaction, pleasure, joy? Every now and then listening to the radio I hear a piece about someone who has been diagnosed as having only three months to live. After the initial shock and grief, their life takes on an immediacy and a clarity. Each day is filled with acts that bring more fullness into their relationships. Every moment is seen to be irreplaceable. They sing for the joy of singing—softly or at the top of their lungs, freed from what other people might think. They write letters, saying what they truly mean. They chop tomatoes and press garlic and delight in eating slowly. They buy fresh flowers and tenderly arrange them. They start a patchwork quilt, for someone else to finish later. They give smiles and hugs in profusion. They enjoy the sweetness of natural light as it plays on grass, trees, and sky through the turning of the day.

For a while after I hear these stories I am more tuned in to my own senses, desires, and urgings from the heart. I let go of goods and objects and take pleasure in the newly found space. Over time I forget and get wound up again in pursuit of schedules and rewards. Then something reminds me I can slow down and start fresh. No matter how disappointed I am that I lost my balance, fell back a step, or hurt someone's feelings, I find relief in remembering that today is a new day and new choices abound. If I am sincere, others will forgive me. And I can learn to forgive myself.

> *What is "enough" in your life? What three things could you do to save money and resources?*

Questions For Reflection

- List ten things you're grateful for. Name ten things you love. How many can you enjoy only with money? How many are unavailable because of the need to work long hours at your current job? How much of your time goes to your passions and priorities?
- How would your life look if you simplified some parts of it? What would you most like to simplify? What three steps can you take in the next week or month to start the process?

- Get three big boxes. Label them Give Away, Recycle, Garbage. Go through one room per week in your house. Notice if a lightening feeling grows as the boxes start to move out.

Chapter 17

What Is and What Can Be

Ellen Schwartz

starting from wherever you are with a passionate consciousness

The best time to plant a tree was 20 years ago. The second best time is now.
—Chinese proverb

Whatever you can do—or dream you can—begin it. Boldness has genius, power, and magic in it. Begin it now!
—Goethe

What did we intend to convey in writing this book?

That things are worse than you think. That we are being betrayed by our democracy. That the media with its polished images is stealing our time and energy and filling our children with self-centered and trivial thoughts. That corporate executives and shareholders are making a killing while 80% of Americans are bringing home less pay and working longer hours than twenty-five years ago. That the frenetic pace of our lives and the fierceness of the competitive mode are debilitating us, physically and spiritually. That economic instability is destroying our families, causing increased verbal dissension and domestic abuse. That our political system is attuned to moneyed interests, not the needs of the general populace. That our educational system is gasping for lack of monetary support and adult involvement. That community is missing for too many people.

Are all these things true for all people? Of course not. But it is the trends in this nation that concern us. We live in a very different country than that of the late 1960s and early 1970s. Deep down we are still an optimistic, can-do, generous people, but the chips are stacked against us with a force and power that we have never before experienced.

It used to be easy to figure out what or who "the enemy" was. The Depression. Communism. Polio. Now it is our very way of life that is destroying us, beguilingly packaged by the transnational corporations who pay for only those media images that forward their purposes while pitting worker against worker with ever smaller paychecks and benefits.

Does this reality check leave us feeling cynical, depressed, hopeless? Not at all! The Chinese character for "crisis" is a combination of "danger" and "opportunity." The water beginning to boil in the cauldron is the perfect cue for us to jump out! But where shall we go? Inward, to the deep, fertile quiet. Outward, to the aliveness of community.

How do we start? By letting go of what never worked! By letting ourselves be seduced by the positive. By filling ourselves and the young people in our lives with the real stories of our struggles and triumphs. By sensing the grace that surrounds and supports us in every moment. By reveling in the awareness that together we are whole.

Some days it's easier than others. One thing we have noticed is that the quality of one day spills over into the next. The sun may rise with fresh radiance, but our spirits carry yesterday's residue. If we have overeaten, we wake up still craving. If we have drunk too much alcohol or taken too many painkillers, our minds and senses are dulled.

In simple gestures we find salvation. The more we give, the more sweetness comes back to caress us.

I remember my dad telling me a story when I was a child. I had a very bad asthma attack, and as we were waiting those grueling minutes for the medication to take hold, the only thing that helped me slow and deepen my breathing was to have Dad tell me a story in a low voice.

When he was a young man during the days of the Depression, he was lucky enough to get a job with the Works Progress Administration (WPA). A crew of men met on the edge of town, and were driven deep into the woods to cut trees all day. One day at noon, when everyone stopped to eat their lunch perched on tree stumps, Dad noticed one fellow sitting with nothing on his lap. "Where's your lunch?" he asked. "Forgot it today" was the reply. Dad had just unwrapped his ham sandwich, cut his apple, and poured a cup of coffee from his thermos. Immediately, he divided them in half. "Really, I couldn't take your lunch," the man protested. "Of course you can. You must. I insist." As I lay there, my breath finally starting to quiet down, Dad said, "I felt as full as if I had eaten the whole sandwich."

In a shared sandwich lies the secret of life.

Afterword: Way More Fun Than TV

surefire ways to release your playful spirit

To live is good. To live vividly is better. To live vividly together is best.
—Anonymous

From our earliest days, we have loved to make lists! Following is a list of our family and friends' favorite things to do. We trust it will pique your imagination and memories of things long forgotten that make life fun all over again.

1. Make a birthday card. Tell the person something you really like about them, or some reason you're proud of them. Make a wonderful wish for them. Draw beautiful flowers or funny faces. Use images cut from magazines or wrapping paper. Make a simple border repeating a design. Go to an art store and buy a big set of colored pencils. You'll marvel at what you can do, blending and contrasting the many hues.
2. Grow tomatoes and sunflowers.
3. Write a journal.
4. Jump rope.
5. Build a doll house. Make the furniture too.
6. Get a big washing machine box to make a secret house or puppet theatre. Paint the box. Paste favorite pictures on it. Make puppets.
7. Read a chapter of J.R.R. Tolkien's *Hobbit* trilogy aloud to the family each night after supper.
8. Get more sleep! Research shows we are generally a sleep-deprived culture. When people get more sleep they feel happier, less anxious and irritable, make fewer mistakes, and are more creative and healthy. Especially during the

winter, our systems have evolved over the millenia to need more deep rest.

9. Take a class you've been wanting to explore. Perhaps you'll meet a new friend.
10. Plan a bike ride with a backpack lunch. Remember crumbs for the ducks and birds.
11. Find out the schedule for the local pool. Dive in!
12. Go to a craft store and lose yourself wandering around.
13. Endear yourself to someone by giving a foot or back massage.
14. Wash the car inside and out.
15. Bring someone you love a cooked meal. Bring someone you don't like a cooked meal, and sit down and get to know them.
16. Take care of a friend's children for a weekend, so they can get a break and the kids can, too!
17. Build a sand castle.
18. Collect quotes and write them on beautiful paper. Put one in front of each person's plate at mealtimes, and have each person talk about what it means to them or what other thoughts it generates.
19. Develop your own new traditions and rituals.
20. Write a poem.
21. Buy and play *The Ungame* and *Life Stories.* As you go around the board in any direction, the cards you pick have tantalizing questions

we don't often get the chance to slip into everyday conversation.

22. Write a story and illustrate it. Make borders around the pages.
23. Get candles of different sizes and press designs into them using thin sheets of colored wax.
24. Make play food from salt-and-flour dough. Paint it with bright acrylics.
25. Walk in nature after supper. Feel the wind in your hair, listen to the twilight calls of the birds, notice the deepening shades of green, enjoy the feeling of your strong hand wrapping around your child's.
26. Bake bread.
27. Meditate.
28. Cut figures out of wood. Paint them and put them on your wall, in your front yard, or give them as presents.
29. Make big butterflies. Make vivid flowers and bright hearts and glittering stars. Hang them from strings strung between corners and across your ceiling.
30. Climb a tree.
31. Build a treehouse or fort.
32. Make salt-water taffy or fudge.
33. Try a new ethnic restaurant.
34. Visit a sick neighbor.
35. Get to know an older couple on your block. Ask them to tell you stories from their child-

hood. Ask them what life was like before television.

36. Make sand candles at the beach.
37. Build a bookcase.
38. Start a book group.
39. Try out for a play.
40. Buy dress-up clothes at the thrift store.
41. Learn folk dances from your favorite culture.
42. Make papier mache figures and bowls.
43. Sit in a garden.
44. Plan a summer solstice gathering.
45. Write a letter to the President, your Representative or Senator.
46. Make a collage for a dream you have or a goal you're trying to achieve. Use images from calendars, greeting cards, magazines, wrapping papers, and your own original drawing.
47. Go on a historical excursion.
48. Make up a play. Take it to a nursing home.
49. Clean out the garage.
50. Go wild making a Halloween costume.
51. Make a gingerbread house or castle. While you're munching the candy, remember to save some for the house.
52. Call a friend you haven't talked to in years.
53. Learn a musical instrument.
54. Join a singing group.
55. Cook a campfire dinner and watch the stars come out.

56. Make seasonal wreaths for your front door. Use dried and silk flowers, herbs, pine cones, ribbons, old jewelry, painted branches, fabric snippets, and whatever else catches your fancy.
57. Sew clothes for baby dolls and bears.
58. Find every available body in the neighborhood of any age and go down to the schoolyard for a baseball game.
59. Make a new recipe for supper.
60. Buy a magnifying glass and examine everything inside and outside your house.
61. Organize the photographs in the shoe boxes.
62. Visit the zoo.
63. Shoot hoops.
64. Visit the library. Check out a pile of books and find out about upcoming special programs.
65. Plan a slumber party.
66. Find out about nearby regional parks.
67. Play charades.
68. Have a mother-daughter tea party.
69. Learn yoga or T'ai Chi.
70. Have a dinner party where everyone brings a passage from their favorite author, or something they've written themselves.
71. Institute a family night. Each family member takes turns picking the meal and deciding what activity everyone will be doing for the evening.
72. Take close-up photographs.
73. Go to a museum.
74. Organize a neighborhood scavenger hunt.

75. Play chess or checkers.
76. Become a tutor.
77. Clean your closet. Lighten up by giving away surplus items to someone who needs them more than you do.
78. Play with someone's dog.
79. Make talking sticks with driftwood, shells, ribbons, and beloved objects.
80. Bake cookies. Make homemade jam.
81. Go ice skating or rollerblading.
82. Paint a mural on your wall.
83. Gather some friends and a big picnic basket full of goodies and go to a storytelling festival.

Love to do something that's not on this list? Write or e-mail us at ssunicorn@earthlink.net with your favorite activities to add to the next printing!

Bibliography

Alterman, Eric. *Sound and Fury: The Washington Punditocracy and the Collapse of American Politics.* New York: HarperCollins, 1992.

Bagdikian, Ben H. *The Media Monopoly.* Boston: Beacon Press, 1992.

Bartlett, Donald L., and James B. Steele. Corporate Welfare (5-part series), *Time,* November 9, 1998.

Benedict Jeff, and Don Yaeger. *Pros and Cons: The Criminals Who Play in the NFL.* New York: Warner Books, 1998.

Canada, Geoffrey. *Fist Stick Knife Gun.* Boston: Beacon Press, 1995.

Covey, Stephen R. *The 7 Habits of Highly Effective People.* New York: Fireside, 1989.

Csikszentmihalyi, Mihaly. *Flow.* New York: Harper Perennial, 1990.

Csikszentmihalyi, Mihaly. *The Evolving Self.* New York: HarperCollins, 1993.

Easwaran, Eknath. *Gandhi the Man.* Berkeley, CA: Blue Mountain Center of Meditation, 1978.

Elkind, David. *The Hurried Child.* Reading, MA: Addison-Wesley, 1981.

Ellul, Jacques. *The Technological Society.* New York: Vintage, 1964.

Foreman, George. *By George: The Autobiography of George Foreman.* New York: Random House, 1998.

Fox, Matthew. *The Reinvention of Work.* San Francisco: HarperSanFrancisco, 1994.

Gibran, Kahlil. *The Prophet.* New York: Knopf, 1967.

Gottlieb, Sanford. *Defense Addiction: Can America Kick the Habit?* Boulder: Westview, 1977.

Greider, William. *Who Will Tell the People? The Betrayal of American Democracy.* New York: Simon & Schuster, 1992.

Grossman, Richard L., and Frank T. Adams. *Taking Care of Business: Citizenship and the Charter of Incorporation* (booklet). Cambridge, MA: Charter, Inc., 1997.

Hawken, Paul. *The Ecology of Commerce.* New York: HarperCollins, 1993.

Hillman, James, and Michael Ventura. *We've had 100 Years of Psychotherapy—and the World's Getting Worse.* San Francisco: HarperSanFrancisco, 1992.

Hollander, Ann. *Seeing Through Clothes.* New York: Viking Penguin, 1988.

Jhally, Sut. *Dreamworlds* (video documentary). Northampton, MA: Media Education Foundation, 1995.

Jhally, Sut. *The Killing Screens* (video documentary). Northampton, MA: Media Education Foundation, 1994.

Kilbourne, Jean. *Feminist Perspectives on Eating Disorders.* Edited by Fallon, Katzman, and Wooley. New York: Guilford Publishing, 1984.

Kohn, Alfie. *No Contest: The Case Against Competition.* Boston: Houghton Mifflin, 1986.

Korten, David. *When Corporations Ruled the World.* San Francisco: Berrett-Koehler, 1995.

Kumar, Satish. *Gandhi's Swadeshi: The Economics of Permanence,* in *The Case Against the Global Economy,* edited by Jerry Mander and Edward Goldsmith. San Francisco, Sierra Club, 1996.

Lappe, Marc, and Britt Bailey. *Against the Grain: Biotechnology and the Corporate Takeover of Your Food.* Monroe, ME: Common Courage, 1998.

Mander, Jerry. *Four Arguments for the Elimination of Television.* New York: Morrow Quill, 1978.

Mander, Jerry. *In The Absence of The Sacred: The Failure of Technology and the Survival of the Indian Nations.* San Francisco: Sierra Club, 1991.

Mander, Jerry, and Edward Goldsmith, editors. *The Case Against the Global Economy.* San Francisco: Sierra Club, 1996.

Mead, Margaret, editor. *Cooperation and Competition Among Primitive Peoples.* Boston: Beacon, 1961 (originally 1937).

Mishel, Lawrence, Jared Bernstein, and John Schmitt. *The State of Working America.* Ithaca, NY: Cornell University Press, 1999.

Myss, Caroline. *Anatomy of the Spirit.* New York: Crown, 1996.

Norberg-Hodge, Helena, and Peter Matthiessen. *Ancient Futures: Learning from Ladakh.* San Francisco: Sierra Club, 1992.

Parenti, Michael. *Democracy for the Few.* New York: St. Martin's, 1995.

Parenti, Michael. *Dirty Truths.* San Francisco: City Lights, 1996.

Pipher, Mary. *The Shelter of Each Other.* New York: Ballantine, 1996.

Postman, Neil. *Amusing Ourselves to Death.* New York: Penguin, 1985.

Project Censored. *Censored 1999.* New York: Seven Stories Press, 1999.

Rifkin, Jeremy. *The End of Work.* New York: Tarcher/Putnam, 1995.

Servan-Schreiber, Jean-Jacques. Reading, MA: *The Art of Time.* Addison-Wesley, 1988.

Shuman, Michael. *Going Local.* New York: Free Press, 1998.

Solomon, Norman, and Jeff Cohen. *Wizards of Media Oz.* Monroe, ME: Common Courage, 1997.

Stauber, John, and Sheldon Rampton. *Toxic Sludge Is Good for You.* Monroe, ME: Common Courage, 1995.

Toms, Michael and Justine. *True Work.* New York: Bell Tower/Crown, 1998.

Walsh, David. *Selling Out America's Children.* Minneapolis: Deaconess, 1994.

Weisman, Alan. *Gaviotas.* White River Junction, VT: Chelsea Green, 1998.

Wheatley, Margaret J., and Myron Kellner-Rogers. *A Simpler Way.* San Francisco: Berrett-Koehler, 1996.

Wolf, Naomi. *The Beauty Myth: How Images of Beauty are Used Against Women.* New York: Morrow, 1991.

Endnotes

Chapter 2. The Gift That Keeps on Taking

[1] Jerry Mander. *In the Absence of the Sacred,* 123-24.
[2] Ibid., 126.
[3] Mander, "The Rules of Corporate Behavior" in *The Case Against The Global Economy,* 310.
[4] Lawrence Mishel, Jared Bernstein, and John Schmitt. *The State of Working America,* 6.
[5] Ibid., 339.
[6] Mishel, Bernstein, and Schmitt. "Finally, Real Wage Gains," Economic Policy Institute Brief #127.
[7] *The State of Working America,* 136-37.
[8] Ibid., 9.
[9] David Korten. "The Failure of Bretton Woods" in *The Case Against the Global Economy,* 26.
[10] Andy Dworkin. "Job Trims Hit Decade High in 98; Continued Mergers May Mean More Cuts." *The Dallas Morning News,* January 9, 1999.
[11] Jeremy Rifkin. *The End of Work,* 138.
[12] Sara Anderson and John Cavanaugh. "The Top 200: The Rise of Global Corporate Power," a report of the Institute for Policy Studies, September 25, 1996.

[13] Kai Mander and Alex Boston. "Wal-Mart: Global Retailer" in *The Case Against the Global Economy,* 336, 339.

[14] Ibid., 342.

[15] Ted Halstead and Clifford Cobb. "If the GDP is Up, Why Is America Down?" *Atlantic Monthly,* October 1995, 59.

[16] Halstead and Cobb. "The Need for a New Measurement of Progress" in *The Case Against The Global Economy,* 202, 204.

[17] Herman E. Daly. "Sustainable Growth? No Thank You" in *The Case Against the Global Economy,* 193.

[18] Marc Lappe and Britt Bailey. *Against the Grain: Biotechnology and the Corporate Takeover of Your Food,* 53-54.

[19] Glen Martin. "Gene-Spliced Corn Imperils Butterfly." *San Francisco Chronicle,* May 20, 1999.

[20] Richard Grossman and Frank Adams. *Taking Care of Business—Citizenship and the Charter of Incorporation.*

[21] Eknath Easwaran. *Gandhi the Man,* 20.

[22] Satish Kumar. "Gandhi's Swadeshi: The Economics of Permanence" in *The Case Against The Global Economy,* 423.

Chapter 3. The Hidden Costs of Competition

[23] Alfie Kohn. *No Contest: The Case Against Competition,* 37.

[24] Ibid., 36.

[25] Ibid., 2.

[26] David Elkind. *The Hurried Child,* 193.

[27] Bob Wieder. "Guns Are for Shooting," *San Francisco Chronicle,* May 11, 1999.

[28] David Walsh. *Selling Out America's Children,* 88.

[29] Jeff Benedict. *Pros and Cons: The Criminals Who Play in the NFL.*

[30] Noam Chomsky. *Manufacturing Consent* (video).

[31] *No Contest: The Case Against Competition,* 25.

[32] Ibid., 72.

[33] Ibid., 90.

[34] Ibid., 102, 103.

[35] Alfie Kohn. Presentation at conference titled "Beyond Conflict: Transcending Us vs. Them." Washington, DC, June 24, 1989.

[36] Stephen Covey. *The 7 Habits of Highly Effective People,* 262-63.

Chapter 4. This Is Entertainment?

[37] Daniel Schorr. 200,000 Acts of Violence, *Christian Science Monitor,* 9/7/93.
[38] Paul Farhi. "Study Finds Violence All Over TV," *Washington Post,* February 6, 1996.
[39] Jerry Mander. *Four Arguments for the Elimination of Television,* 180.
[40] Neil Postman. *Amusing Ourselves to Death,* 8.
[41] David Walsh. *Selling Out America's Children,* 52-54.
[42] Ben Bagdikian. *The Media Monopoly,* xxvii.
[43] Norman Solomon and Jeff Cohen. *Wizards of Media Oz,* 82.
[44] Michael Parenti. *Dirty Truths,* 100.
[45] John Stauber and Sheldon Rampton. *Toxic Sludge Is Good for You,* 180.
[46] Ibid., 2.
[47] Ibid., 2.
[48] Ibid., 3, 4.
[49] Ibid., 13.
[50] Ibid., 4.
[51] "Paying the Piper." Friends of the Earth newsletter, Fall 1998, 8.
[52] Media Education Foundation. *The Killing Screens: Media and the Culture of Violence* (video documentary).
[53] Mihaly Csikszentmihalyi. *Flow.*

Chapter 5. Media, Girls, and Body Image

[54] *New Age Journal,* July/August 1996, 21.
[55] Sut Jhally. *Dreamworlds* (documentary video).
[56] "Youth Risk Behavior Survey." The National Center for Health Statistics and the Alan Guttmacher Institute, 1995.
[57] Naomi Wolf. "Hunger." Essay in *Feminist Perspectives on Eating Disorders,* 94.

Chapter 7. What Do World Trade Agreements Have To Do With Me?

[58] Herman Daly. "Free Trade: The Perils of Deregulation" in *The Case Against the Global Economy,* 230.
[59] Ralph Nader and Lori Wallach. "GATT, NAFTA, and the Subversion of the Democratic Process" in *The Case Against the Global Economy,* 104.
[60] Public Citizen's website: www.tradewatch.org.
[61] Greenpeace Report: "Chlorine, Human Health, and the Environment: The Breast Cancer Warning." October 1993.
[62] Comments of Public Citizen, Inc. regarding U.S. Preparations for the World Trade Organization's Ministerial Meeting, October 1998, www.tradewatch.org.

Chapter 13. Meaningful Work

[63] Matthew Fox, *Reinvention of Work,* 89-90.
[64] E.F. Schumacher, *Good Work.*
[65] Paul Hawken. *The Ecology of Commerce,* 62.
[66] Ibid, 63.
[67] Ibid., xiv-xv.
[68] Geoffrey Canada. *Fist Stick Knife Gun,* 160.
[69] Jeremy Rifkin. *The End of Work,* 258.
[70] Donald L. Bartlett and James B. Steele. "Corporate Welfare," 5-part series. *Time,* November 9, 1998.

Chapter 15. Together We Are Whole

[71] Margaret J. Wheatley and Myron Kellner-Rogers. *A Simpler Way,* 33.
[72] Ibid., 39.

If you have any comments you'd like to share with us, please email us at: ssunicorn@earthlink.net or write us at P.O. Box 5667, Walnut Creek, CA 94596.

Resources

Chapter 2—The Gift That Keeps On Taking

Boycott Quarterly, P.O. Box 30727, Seattle, WA 98103, 206-789-8383, e-mail: boycottguy@aol.com .

The Corporate Crime Reporter at corporatepredator.org.

Extra!, the news magazine of Fairness and Accuracy in Reporting, Inc., P.O. Box 170, Congers, NY 10920-9930.

Friends of the Earth's "Green Scissors Report," 1025 Vermont Ave. NW, Suite 300, Washington, DC 20005-6303, 202-783-7400, fax: 202-783-0444, web: www.foe.org.

The Hightower Lowdown at www.JimHightower.com.

How Wal-Mart Is Destroying America: And What You Can Do About It, by Bill Quinn. Berkeley, CA: Ten Speed Press, 1998.

Multinational Monitor, Essential Information, Inc., 1530 P Street NW, Washington, DC 20005, 202-387-8030, monitor@essential.org.

Advocacy Groups

Businesses for Social Responsibility, 1030 15th St. NW, Suite 1010, Washington, DC 20005, 202-842-5400.

Economic Policy Institute, 1660 L St. NW, Suite 1200, Washington, DC 20036, 202-775-8810, fax: 202-775-0819, e-mail: economic@cais.com, web: www.epinet.org.

Environmental Defense Fund. Find out about chemical pollution in your own community by visiting their web site at www.scorecard.org.

Institute for Policy Studies, 733 15th St. NW, Suite 1020, Washington, DC 20005, 202-234-9382, fax: 202-387-7915, e-mail: shuman@igc.org, web site: www.ic.apc.org/ifps.

Interfaith Center on Corporate Responsibility, 212-870-2293, 475 Riverside Drive, Room 550, New York, NY 10115.

International Forum on Globalization, 1555 Pacific Avenue, San Francisco, CA 94109, 415-771-3394, www.ifg.org.

Program on Corporations, Law, and Democracy, P.O. Box 806, Cambridge, MA 02141, phone and fax: 508-487-3151.

Chapter 3—The Hidden Costs of Competition

Print

Fluegelman, Andrew. *The New Games Book.* Garden City, NY: Dolphin, 1976.

Kohn, Alfie. *No Contest: The Case Against Competition.* Boston: Houghton Mifflin, 1986.

Orlick, Terry. *The Second Cooperative Sports and Games Book.* New York: Pantheon, 1982.

Slavin, Robert et al., editors. *Learning to Cooperate, Cooperating to Learn.* New York: Plenum, 1985.

Sobel, Jeffrey. *Everybody Wins.* New York: Walker, 1982.

Tutko, Thomas, and Bruns, William. *Winning is Everything and Other American Myths.* New York: MacMillan, 1986.

Video

Chomsky, Noam. *Manufacturing Consent.* Montreal, Quebec: National Film Board of Canada, 1992.

Chapter 4—This Is Entertainment?

Video

"The Killing Screens" with George Gerbner. Media Education Foundation, 26 Center St., Northampton, MA 01060, 413-586-4170.

Advocacy Groups

TV-Free America, 202-887-0436, web: www.tvfa.org .*Unplug (Channel One),* 360 Grand Avenue, Oakland, CA 94612, 510-268-1100.

Television Stations

Major Networks

ABC-TV, Capital Cities/ABC, 77 W. 66th Street, 9th Floor, New York, NY 10023, 212-456-7777, www.abc.com.

CBS-TV, 51 W. 52nd Street, New York, NY 10019, 212-975-4321, www.cbs.com.

NBC-TV, 30 Rockefeller Plaza, 25th Floor, New York, NY 10112, 212-664-4444, www.nbc.com.

PBS-TV, Elementary/Secondary Service, 1320 Braddock Place, Alexandria, VA 22314, 703-739-5000, www.pbs.org.

CBC-TV, Canadian Broadcasting Corp., Box 500 Sta. A, Toronto, Ontario, Canada M5W 1E6, 416-205-3311, www.tv.cbc.ca.

Cable Channels

A&E (Arts & Entertainment), Viewer Relations, 235 E. 45th St., New York, NY 10017, 212-210-1400 (same address for History Channel).

AMC (American Movie Classics) and BRAVO, 1111 Stewart Ave., Bethpage, NY 11714, 516-396-4500, www.amctv.com.

BET (Black Entertainment Television), 1700 N. Moore St., Suite 2200, Rosslyn, VA 22201, www.cnbc.com.

CNBC, 2200 Fletcher Avenue, Ft. Lee, NJ 07024 (specify Viewer Services), 201-585-2622, www.cnbc.com.

CNN (Cable News Network), TBS Superstation and TNT (Turner Network Television), Turner Educational Services, 1 CNN Center, Atlanta, GA 30348-5366, 404-827-2252, learning.turner.com.

C-SPAN, 400 N. Capitol St., Suite 650, Washington, DC 20001, 202-737-3220, fax: 202-737-3323.

DIS (The Disney Channel), 3800 W. Alameda, Burbank, CA 91505, 818-569-7500.

ESPN Headquarters, Viewer Relations, 860-585-2236.

ESPN, Inc., 605 3rd Ave., New York, NY 10158, 212-916-9200.

FAM (The Family Channel), Viewer Relations, 2877 Guardian Lane, P.O. Box 2050, Virginia Beach, VA 23450-2050, fax 757-459-6420.

HBO (Home Box Office) and MAX (Cinemax), HBO/Cinemax, 1100 Avenue of the Americas, New York, NY 10036, 212-512-1000, www.hbo.com.

LIFE (Lifetime), 3612 35th Avenue, Astoria, NY 11106.

MON (The Monitor Channel), One Norway St., Boston, MA 02115.

NICK (Nickelodeon) and MTV (Music Television), 1515 Broadway, New York, NY 10036, 212-258-7500.

SHOW (Showtime) and TMC (The Movie Channel), 1633 Broadway, New York, NY 10019, 212-708-1600, www.showtime.com.

TDC (The Discovery Channel) and TLC (The Learning Channel), 7700 Wisconsin Ave., Bethesda, MD 20814-3522, 301-986-1999, web: www.discovery.com.

USA (USA Network), 1230 Avenue of the Americas, New York, NY 10020, 212-408-9100, Media Relations: 212-408-8228.

VISN (Vision Interfaith Satellite Network), 74 Trinity Place, Eighth Floor, New York, NY 10006, 212-602-0738.

Check your local newspaper's TV listings for addresses of local TV stations and outlets.

Chapter 5—Media, Girls, and Body Image

Print

New Moon, a magazine for and by girls aged 9 to 14, P.O. Box 3587, Duluth, MN 55803-3587.

Video

Dreamworlds. The Media Foundation, 26 Center St., Northampton, MA 01060, 413-586-4170.

Chapter 6—The Best Government Big Money Can Buy

Alternative Media

Radio

Pacifica Network Affiliates: KPFA Berkeley—94.1, KPFK Los Angeles—90.7, KPFW Washington—89.3, KPFT Houston—90.1, or WBAI New York—99.5.

New Dimensions Radio Network, Michael and Justine Toms, P.O. Box 569, Ukiah, CA 95482, 707-468-5215, e-mail: ndradio@igc.org.

Alternative Political Parties

Green Party: 916-448-3437
Labor Party: 202-234-5190
New Party: 718-246-3713
Natural Law Party: 515-472-2040

Advocacy Groups

Center for Defense Information, 1779 Massachusetts Ave., NW, Washington, D.C.

Center for Living Democracy, 289 Fox Farm Rd., Brattleboro, VT 05301, 802-254-1234, fax: 802-254-1227, e-mail: cld@sover.net, web: www.sover.net/~cld.

Center for Responsive Politics, 1320 19th St., NW, Suite 700, Washington, DC 20036, 202-857-0044, fax: 202-857-7809, e-mail: info@crp.org, web: www.crp.org.

CyberCitizen, www.citizen.org, for exercising citizenship online.

Tri-Valley Communities Against a Radioactive Environment, 925-443-7148, 2582 Old First Street, Livermore, CA 94550, web: www.igc.org/tvc

Government

Library of Congress Thomas web site for information on all pending bills in the Senate and House: thoma s.loc.gov.

Government Addresses:

Your Representative
U. S. House of Representatives
Washington, DC 20515
Your Senator
U. S. Senate
Washington, DC 20510

For local and state elected officials, see the front of your phone book.

To download copies of State Senate bills in California, use the web site www.sen.ca.gov.

Chapter 7—What Do World Trade Agreements Have to Do with Me?

Advocacy Groups

Ad Hoc Working Group on the MAI, 1-800-316-APEX.

Breast Cancer Action, 415-243-8373.

Fair Trade Foundation, 65 Landing Rd., Hagganum, CT 06441, 860-345-3374, fax: 860-345-4922.

Public Citizen's Global Trade Watch at 202-546-4996, or look up their web site at www.citizen.org.

Public Citizen News, 1600 20th St., NW, Washington, DC 20009, www.citizen.org.

Video

Who's Counting? Marilyn Waring on Sex, Lies, and Global Economics. Bullfrog Films, P.O. Box 149, Oley, PA 19547, 610-779-8226, e-mail: bullfrog@igc.apc.org.

Chapter 10—What's An Inner Life and Who Needs It?

Print

Franquemont, Sharon. *You Already Know What to Do: 10 Invitations to the Intuitive Life.* New York: Tarcher/Putnam, 1999.

Spiritual Centers

Nyingma Institute, 1815 Highland Place, Berkeley, CA 94709-1009, 510-843-6812.

Self-Realization Fellowship, 6401 Bernhard Ave., Richmond, CA 94805-1655, 510-234-2494.

Siddha Yoga Ashram, 1107 Stanford Ave., Oakland, CA 94608, 510-655-8677.

Chapter 12—Nurturing What Is Precious

Print

Montagu, Ashley. *Touching: The Human Significance of the Skin.* New York: Harper & Row, 1978.

Chapter 13—Meaningful Work

Print

Shuman, Michael. *Going Local: Creating Self Reliant Communities in a Global Age.* New York: Free Press, 1998.

Schumacher, E.F. *Small Is Beautiful: Economics as if People Mattered.* New York: Harper Collins, 1973.

YES! A Journal of Positive Futures, P.O. Box 10818, Bainbridge Island, WA 98110-0818, 206-842-0216, fax: 206-842-5208, e-mail: yes@ futurenet.org, web: www.futurenet.org.

Advocacy Groups

The Center on Budget and Policy Priorities, 202-408-1080.

Community Jobs, 50 Beacon St., Boston, MA 02108, 617-720-5627.

Job Seeker, 28672 County EW, Warrens, WI 54666, phone and fax: 608-378-4290, e-mail: jo bseeker@tomah.com, web: www.tomah.com/job seeker.

National Center for Employee Ownership, 1201 Martin Luther King Jr. Way, Oakland, CA 94612, 510-272-9461, fax: 510-272-9510, e-mail: nceo@nceo.org, web: www.nceo.org.

New Road Map Foundation, P.O. Box 15981, Seattle, WA 98115, 206-527-0437, fax: 206-528-1120.

New Ways to Work, 785 Market Street, San Francisco, CA, 415-995-9860.

Planet Drum Foundation, creating self-sustaining economies within a geographic region, P.O. Box 31251, San Francisco, CA 94131, 415-285-6556, fax: 415-285-6563, e-mail: planetdrum@igc.org.

Real Goods, earth-friendly technologies and products, 200 Clara Ave., Ukiah, CA. 95482, 800-762-7352, www.realgoods.com

The Rocky Mountain Institute (founded by Amory Lovins), 1739 Snowmass Creek Rd., Snowmass, CO 81654-9199, 970-927-3851, www.rmi.org.

Chapter 14—Giving Time, Getting Joy

Print

Boal, John T. *Be a Global Force of One! ... In Your Hometown.* Burbank: PacRim, 1999.

Advocacy Groups

Earthwatch Institute, 680 Mt. Auburn St., P.O. Box 9104, Watertown, MA 02272-9104, 800-776-0188, e-mail: info@earthwatch.org,www.earthwatch.org.

Global Exchange, 2017 Mission St., Suite 303, San Francisco, CA 94110, 415-255-7296.

Global Service Corp., Suite 28, 300 Broadway, San Francisco, CA 94133-3312, 415-788-3666, Ext.128, e-mail: gsc@igc.apc.org.

Global Volunteers, 375 E. Little Canada Road, St. Paul, MN 55117, 800-487-1074.

Head Start (national organization), 703-739-0875.

Heifer Project International, Route 2, Box 33, Perryville, AR 72126, 501-889-5124.

Invest in Kids mentoring and scholarship program, P.O. Box 5667, Walnut Creek, CA 94596, 925-932-6943.

Institute for Food and Development Policy (Food First), 398 60th St., Oakland, CA 94618, 510-654-4400, fax: 510-654-4551, e-mail: foodfirst @igc.org, web: www.netspace.org/hungerweb/Food First/index.htm.

Oceanic Society Expeditions, Suite 230, Building E, Fort Mason Center, San Francisco, CA 94123, 800-326-7491.

Pesticide Action Network, 415-541-9140 or 415-981-1606, 49 Powell St., San Francisco, CA.

Volunteer Center of ___________ County. (Look in the phone book for your area.)

Volunteers for Peace, 43 Tiffany Road, Belmont, VT 05730, 802-259-2759, e-mail: vfp@vermontel.com.

Video

Mentor (CD), Bay Area Songwriters, Richmond, CA: Daniel-Hill-Daniel, 1999.

Chapter 15—Together We Are Whole

Print

Cohousing: Journal of the Cohousing Network, P.O. Box 2584, Berkeley, CA 94702, 510-486-2656, email: cohomag@aol.com, web: www.cohousing.org.

Macy, Joanna, and Molly Young Brown. *Coming Back to Life: Practices to Reconnect Our Lives, Our World.* Gabriola Island, BC: New Society, 1998.

Communities Magazine, P.O. Box 160, Masonville, CO 80541, 970-593-5615, e-mail: communities@ic.org, web: www.ic.org/fic/cmag/index.html.

Shaffer, Carolyn, and Kristin Anundsen. *Creating Community Anywhere.* New York: Tarcher/Perigee, 1993.

YES! A Journal of Positive Futures, P.O. Box 10818, Bainbridge Island, WA 98110-0818, 206-842-0216, fax: 206-842-5208, e-mail: yes@futurenet.org, web: www.futurenet.org.

Advocacy Groups

Cohousing Company, 1250 Addison Street, #113, Berkeley, CA 94702, 510-549-9980.

Cohousing Network, Berkeley, California, 510-486-2656.

Community Building Exercises, Sirius Community, P.O. Box 388, Amherst, MA 01004.

Co-op America, 1612 K St., NW, Suite 600, Washington, DC 20006, 800-58-GREEN, fax: 202-331-8166, e-mail: info@coopamerica.org, web: www.coopamerica.org.

Ecovillage at Ithaca, Cornell University, Anabel Taylor Hall, Ithaca, NY 14853,607-255-8276, fax: 607-255-9985, e-mail: ecovillage@cornell.edu, web: www.cfe.cornell.edu/ecovillage/evi.html.

FriendlyFavors.org, 3800 Vista Oaks Dr., Suite 210, Martinez, CA 94553, 925-229-3600, Ext.113, fax: 925-229-3609, e-mail: favors@sergiolub.com.

Institute for Local Self-Reliance, 2425 18th St., NW, Washington, DC 20009, 202-232-4108, fax: 202-332-0463, web: www.ilsr.org.

Neighborhood Salon Association, c/o Utne Reader, P.O. Box 7460, Red Oak, IA 51591, 800-736-UTNE, fax: 612-338-6043, info@utne.com, web: www.utne .com.

Sustainable America, 350 Fifth Ave., Room 3112, New York, NY 10118-3199, 212-239-4221, fax: 212-239-3670, e-mail: sustamer@igc.org, web: ww w.sanetwork.org.

About the Authors

Ellen Schwartz and **Suzanne Stoddard** never take on small challenges. Their shared zest for life and for making a difference have brought them together on a path of personal, social, and spiritual transformation. Schwartz's experience in the political and nonprofit worlds combined with Stoddard's corporate and public-school teaching background offer a compelling view of the stresses and potentialities of life on the cusp of the new millenium.

Schwartz is the executive director of *Invest in Kids,* which she founded in 1998 to provide scholarships and mentors to low-income, at-risk youth. An earlier nonprofit she created raised public awareness about the entertainment industry's negative influence in shaping children's desires, behaviors, and values. Schwartz has spoken in over 160 venues on these issues and been interviewed for dozens of newspaper and magazine articles, radio shows, and television programs. For this work, she won the *Thread of Hope Award* in 1996, and the *Women of Achievement Community and Human Services Award* in 1997.

In her run for the U.S. House of Representatives in 1994, Schwartz received the most votes of any candidate on the West Coast running against an incumbent. Earlier, as an entrepreneur starting and running a rental company with her husband, the welfare of her 27 employees was always given top consideration in decisions about growing the business.

Her commitment to a sustainable environment led her to play a major role in organizing Earth Day Exhibitions in the early 1990s. While the Cold War still raged in the mid-1980s, Schwartz was a co-founder of *The Center for US – USSR Initiatives,* bringing Soviets and Americans together to shatter stereotypes and build friendships and commercial exchanges. Schwartz earned a master's degree in speech and communication studies from San Francisco State University in 1983. She has been married for 23 years, and has a son, 22, and a daughter, 13.

Stoddard's high-energy approach to creating community has included team-building in the corporate realm and donating numerous hours as a tutor, mentor to young people, and board member of *Invest in Kids.* An English and citizenship instructor at a community college, Stoddard worked as a marketing consultant at Pacific Gas and Electric from 1990 to 1996, where she founded and edited an inter-utility newsletter on energy conservation and promoted volunteerism through employee channels. From 1988 to 1990, Stoddard facilitated workshops in the former Soviet Union for Russian and Ukrainian teachers of English, and established penpal relationships between Russian and American children.

Stoddard was active in the 1980s in peace and antinuclear weapons work, speaking widely throughout the San Francisco Bay Area, organizing educational events, and co-founding a bi-monthly newspaper, *The Turning Point,* devoted to these issues. Stoddard has been a volunteer in national and local political campaigns since she was 13 years old. Stoddard has a master's degree in English from San Francisco State University. She has been married for 24 years.

A

B

C

D

E

F

G

H

I

N

T

Y

Z

Books For ALL Kinds of Readers

Made in United States
Orlando, FL
04 November 2026

90202357R00188